Power Systems

More information about this series at http://www.springer.com/series/4622

Tharangika Bambaravanage
Asanka Rodrigo · Sisil Kumarawadu

Modeling, Simulation, and Control of a Medium-Scale Power System

 Springer

Tharangika Bambaravanage
Department of Electrical Engineering
University of Moratuwa
Moratuwa
Sri Lanka

Sisil Kumarawadu
Department of Electrical Engineering
University of Moratuwa
Moratuwa
Sri Lanka

Asanka Rodrigo
Department of Electrical Engineering
University of Moratuwa
Moratuwa
Sri Lanka

ISSN 1612-1287   ISSN 1860-4676   (electronic)
Power Systems
ISBN 978-981-13-5263-8   ISBN 978-981-10-4910-1   (eBook)
https://doi.org/10.1007/978-981-10-4910-1

Printed on acid-free paper

This Springer imprint is published by Springer Nature
The registered company is Springer Nature Singapore Pte Ltd.
The registered company address is: 152 Beach Road, #21-01/04 Gateway East, Singapore 189721, Singapore

To
Mr. D.S.C. Wijesekara
&
Mrs. D.W. Jayasinghe

Acknowledgements

We are very thankful to the Institute of Technology, University of Moratuwa, for the financial support provided for this research study.

Heartfelt sincere thanks are due to Dr. (Miss) Lidula Vidanagamage for help with PSCAD simulations during the course of this work.

We would also like to extend our sincere thanks to the faculty and the staff of the Department of Electrical Engineering, University of Moratuwa, for providing resources and facilitating the research study.

We are very much thankful to the System Control Centre of the Ceylon Electricity Board (CEB), for the invaluable discussions and help with data during this research study.

Last but not least, a precious thank-you goes to our family members for their love, encouragement, and constant support all throughout.

Contents

Contents xi

About the Authors

Tharangika Bambaravanage obtained her B.Sc.(Engineering), M.Eng. in Electrical Engineering, and M. Phil. in Power System Stability and Control from the University of Moratuwa, respectively, in 1998, 2005, and 2017. She has been a Senior Lecturer in Electrical Engineering at the Institute of Technology, University of Moratuwa, since 2017.

Asanka Rodrigo obtained his B.Sc.(Hons) and M.Sc. in Electrical Engineering, respectively, in 2002 and 2004, from the University of Moratuwa, and Ph.D. in Industrial Engineering from Hong Kong University of Science and Technology in 2010. He has been a Senior Lecturer in Electrical Engineering at the Faculty of Engineering of University of Moratuwa since 2010.

Sisil Kumarawadu obtained his B.Sc.(Hons) in Electrical Engineering from the University of Moratuwa in 1996. He obtained his M.Eng. in advanced Systems Control Engineering and Ph.D. in Robotics and Intelligent Systems in 2000 and 2003, respectively, from Saga National University, Japan. From April 2003 to July 2005, he was with Intelligent Transportation Systems Research Center, NCTU, Taiwan, as a postdoctoral research fellow. Currently, he is a Professor in Electrical Engineering at the University of Moratuwa.

List of Figures

List of Tables

Abstract

Emergency load shedding for preventing frequency degradation is an established practice all over the world. The objective of load shedding is to balance load and generation of a particular Power System (PS). In addition to the hydro and thermal generators each with less than 100 MW, today, the PS of Sri Lanka is comprised of three coal power plants: Each has a generation capacity of 300 MW; Yugadanavi combined cycle power plant (300 MW generation capacity) and a considerably extended transmission network. To cater consumers with high-quality electricity, a reliable PS is a must. Therefore, it has become timely necessity to review the performance of the present Ceylon Electricity Board (CEB) Under Frequency Load Shedding Scheme (UFLSS) and suggest amendments where necessary.

The objective of this research is to explore a better UFLSS which can face probable contingencies and maintain stability of the system while catering more consumers. The suggested UFLSSs can address the recent changes taken place in the Sri Lanka PS too.

A simulation of the PS of Sri Lanka was designed with software PSCAD. Its validity was checked through implementing actual scenarios which took place in the PS under approximately equal loaded conditions and by comparing the simulated results and actual results. Then, a performance analysis was done for the CEB UFLSS which is being implemented in Sri Lanka. Having identified its drawbacks, the new UFLSSs (LSS-I and LSS-II) were suggested.

The Load Shedding Scheme-I (LSS-I) is designed based on PS frequency and its derivative under abnormal conditions. Without doing much modification to the prevailing UFLSS, and utilizing the available resources, the suggested LSS-I can be implemented.

The LSS-II gives priority for 40% of the system load for continuous power supply, and it is comprised of two stages. During the stage-I, approximately 30% of the load is involved with the Load Shedding action. During the stage-II, the disintegration of the PS is done. This involves the balance 30% of the load. At 48.6 Hz, the disintegration of the PS takes place. By disintegrating the PS at the above-mentioned frequency, all islands as well as the national grid can be brought to steady-state condition without violating the stability constraints of the Sri Lanka PS. During disintegration of the PS, special attention must be paid for:

- Generation and load balance in each island and in the national grid.
- Reactive power compensation in islands and in the national grid.
- Tripping off of all isolated transmission lines (which are not connected to effective loads).

Through simulations, the effectiveness of the UFLSSs was evaluated. They demonstrate better performance compared to that of the currently implementing CEB scheme. Results highlight that the UFLSS should exclusively be specific for a particular PS. It depends on factors such as PS practice, PS regulations, largest generator capacity, electricity consumption pattern etc.

Chapter 1
Introduction

Both electric utilities and electricity consumers are becoming increasingly concerned about the quality and reliability of electric power while having a healthy (stable) power system.

- Newer-generation load equipment, with microprocessor-based controls and power electronic devices, is more sensitive to power quality variations than the equipment used in the past.
- The increasing emphasis on overall power system efficiency has resulted in continued growth in the application of devices such as high-efficiency, adjustable-speed motor drives and shunt capacitors for power factor correction to reduce losses.
- Consumers have an increased awareness of power quality and reliability issues. They are becoming better informed about such issues as interruptions, sags, and switching transients.
- Many things are interconnected in a network. Hence a failure of any component in the integrated processes or system, can have a considerable impact on the system itself [1].

As a developing country, still Sri Lanka possesses a small power system with an average maximum demand of 2100 MW that occurs around 8.30 p.m. As a result of the word wide development in the microprocessor-based controls and power electronic devices, information and communication technologies, the consumers also tend to go for/with them due to many reasons. As explained before, these equipment are more sensitive to power quality variations than the equipment used in the past. As examples:

- A basic laptop computer (may be manufactured in China) can go out of order, if the supply power frequency is not within its rated limits. Since the users are usually office staff, university graduate and under graduate students, people in the research oriented disciplines, entrepreneurs etc., the outcome may be very bad.

© Springer Nature Singapore Pte Ltd. 2018
T. Bambaravanage et al., *Modeling, Simulation, and Control of a Medium-Scale Power System*, Power Systems, https://doi.org/10.1007/978-981-10-4910-1_1

- Various production processes equipped with power electronic applications, may be affected very badly and can worsen this situation. As a solution the consumers can go for equipment such as voltage regulators etc. Due to very high production cost, price of their produce may not be in the affordable range. Therefore it has become a national requirement to facilitate the consumer with a quality and reliable electricity supply so that they can reduce their production cost. This paves the path to help the manufacturers and to develop the production and manufacturing sector in the country.
- There is a trend to generate electricity based on non-conventional and renewable energy sources such as wind, bio-mass, solar power etc. Even though wind and solar are very good sources of energy, they are with high variability and uncertainty. Therefore the power system must be ready to accept electricity based on such sources, when it is available. This causes the system power to move both directions (forward and reverse).

Due to the continued push for increasing productivity, manufacturers want faster, more productive and more efficient machinery. To avoid such situations discussed above where the consumers can get disturbed very badly, the utilities/authorities should concern more about the quality and reliability of electric power. In other words, the economy of a country highly depends on the quality and reliability of that country's electrical power system.

A mismatch between generation and demand can lead the power system unstable [2, 3]. This mismatch may be due to,

- Sudden increase of generation than the load—if the power consumption is much less than the power generation, unless the Automatic Generation Control (AGC) System acts to rapidly reduce generation, there will be a critical situation which may even leads the system to a collapse of the entire generating system.
- Sudden deficit of generation than the load—if the electricity consumption is much higher than the power generation, unless the AGC system acts fast and produce more electricity, there will be an emergency situation, and a failure is possible.

A disturbance (general) can be identified as: 'An undesired variable applied to a system that tends to affect adversely the value of a controlled variable' [4]. It can be categorized into two major categories.
They are:

- Load disturbances

  - Small random fluctuations super imposed on slowly varying loads

- Event disturbances

  - Faults on transmission lines due to equipment malfunctions or natural phenomena such as lightening.
  - Cascading events due to protective relay action following severe overloads or violation of operating limits.

– Generation outages due to loss of synchronism or malfunction.

As explained in [4], load disturbances are a part of the system normal operating conditions. In an operating power system, frequency and voltage are always in a state of change due to load disturbances. Any departure from normal frequency and voltage, due to a load disturbance, is usually small and requires no explicit power plant or protective system response. Occasionally, however, major load disturbances which can be considered as event disturbances do occur. These may cause situations such as:

- Under frequency
- Over frequency
- widely varying frequency
- reduced system voltage
- increased system voltage
- widely varying voltage on the power system.

Generation deficit is one of the main issues that may occur in the power system of Sri Lanka. This can be identified as an event disturbance. Due to generation deficit, the most possible, significant and immediate outcomes are under-frequency situation and reduced-system-voltage situation. Effects of under-frequency on the power system in turn are undesirable. They are:

- generators get over loaded
- speed is below normal
- cooling is below normal
- system voltages are likely to be low
- generator excitations get increased
- increased possibility of thermal over-load of stator and rotor

These effects may lead the unit trip due to:

- stator over heating
- rotor over heating
- over-excitation
- under-frequency (Volts/Hertz)

Further, due to short of generation in the power system, tripping of any unit can start a cascading of unit trips leading to a black-out condition. Even though a sufficient spin-reserve is available, if proper co-ordination is not there in the power system (such as AGC), still it can lead to a cascading of unit trips. An evidence for such a situation is 'a breakdown of a conductor in the switch-yard, in the Kelenitissa power station' that led to a total collapse of the Sri Lanka power system in 2009.

Even though the fault is cleared the power system can't be brought back to normal quickly, rather the system should be brought to normal state integrating the total grid part by part. It is a process which consumes several hours to be back to normal [5]. Hence such situation can disturb the consumer in different ways. Due to loss of power,

- Production of various items can get halted.
- In certain processes not only the production is stopped, the raw materials used also won't be reused.
- Idling of man-power (skilled and unskilled labor).
- Due to delayed supplies other linked projects also get interrupted.

The monitory value of the above outcomes may worth of millions of rupees. Therefore it is essential to be vigilant on the over-all power system and to maintain a quality and reliable power supply to cater the consumers.

Hence, event disturbances, need quick response by the protective systems and can lead to larger upsets if this action fails or is delayed. Large event disturbances always require fast protective system action and may lead to complete system failure if this action is not correct and fast. One of such solutions is to maintain the system stability through introducing an Under Frequency Load Shedding Scheme. Among the factors that affect in designing a suitable load shedding scheme, 'maximum generation capacity of the largest generator unit/s available in the power system' is significant. According to the experience with regard to the black-out situations occurred in the power system of Sri Lanka during the past few years, lack of a proper/updated Under Frequency Load Shedding Scheme was a main reason.

Chapter 2
Literature Survey

2.1 Structure of an Electrical Power System

Electrical power systems are large, complex structures consisting of power sources, transmission networks, distribution networks and a variety of consumers (Fig. 2.1). Only the generation and transmission levels are considered in power system analysis; the distribution networks are not usually modelled as-such, but replaced by equivalent loads—composite loads [6]. **Appendix-A** demonstrates the modeling of the power system's main power corridor—transmission lines and transformers with a two-port network.

With regard to the power system engineering, in general, load can be considered as:

- A device connected to the power system that consumes power;
- The total active or reactive power consumed by all devices connected to the power system;
- The power output of a particular generator plant;
- A portion of the system that is not explicitly represented in the system model, but as if it were a single power-consuming device—composite load [**Appendix-B**].

2.2 Power System Stability

Stability of a power system can be defined as: The ability of an electric power system, for a given initial operating condition, to regain a state of operating equilibrium after being subjected to a physical disturbance, with most system variables bounded so that practically the entire system remains intact [7].

© Springer Nature Singapore Pte Ltd. 2018
T. Bambaravanage et al., *Modeling, Simulation, and Control of a Medium-Scale Power System*, Power Systems, https://doi.org/10.1007/978-981-10-4910-1_2

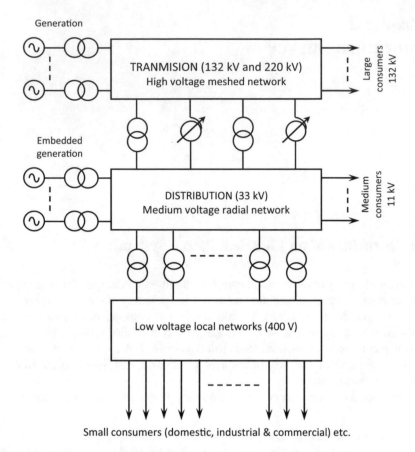

Fig. 2.1 Structure of an electrical power system

2.3   Why Power System Instability Situations Occur?

Power system is a highly nonlinear system that operates in a constantly changing environment. Because of the interconnection of different elements that form a large, complex and dynamic system capable of generating, transmitting and distributing electricity, in a power system, a large variety of dynamic interactions are possible. The main causes of PS dynamics are:

- changing power demand
- various types of disturbances

A changing power demand introduces a wide spectrum of dynamic changes into the system each of which occurs on a different time scale.

- The fastest dynamics are due to sudden changes in demand—which associates with the transfer of energy between the rotating masses in the generators and the loads.
- Slightly slower are the voltage and frequency control actions needed to maintain system operating conditions
- very slow dynamics which corresponds to the way in which the generation is adjusted to meet the slow daily demand variations

The way in which the system responds to disturbances also covers a wide spectrum of dynamics and associated time frames.

- The fastest dynamics are those associated with the very fast wave phenomena that occur in high-voltage transmission lines.
- Slightly slower are the electromagnetic changes in the electrical machines which occur before the relatively slow electromechanical rotor oscillations occur
- The slowest are the prime mover and automatic generation control actions

For reliable service, a PS must remain intact and be capable of withstanding a wide variety of disturbances [6, 8–10].

2.4   Disturbances

The disturbances which are experienced by the Power System may be small or large. They vary in both magnitude and character. A disturbance is defined as 'Disturbance (General): An undesired variable applied to a system that tends to affect adversely the value of a controlled variable' [4]. Therefore, it is essential that the system be designed and operated so that the more probable contingencies can be sustained with no loss of load (except that is connected to the faulted element) and the most adverse possible contingencies do not result in uncontrolled, widespread and cascading power interruptions [3, 11, 12]. Disturbance types and characteristics can be categorized into two major categories. They are:

- Load disturbances

  - Small random fluctuations super imposed on slowly varying loads

- Event disturbances

  - Faults on transmission lines due to equipment malfunctions or natural phenomena such as lightening.
  - Cascading events due to protective relay action following severe overloads or violation of operating limits.
  - Generation outages due to loss of synchronism or malfunction.

2.4.1 Effects of the Disturbances on the Power System

The PS responds to these disturbances with the involvement of different equipment:

- A short circuit on a critical element followed by its isolation by protective relays will cause variations in power transfers, machine rotor speeds, and bus voltages;
- The voltage variations will actuate both generator and transmission system voltage regulators; the speed variations will actuate prime mover governors;
- The change in tie line loadings may actuate generation controls;
- The changes in voltage and frequency will affect loads on the system in varying degrees depending on their individual characteristics.
- Devices used to protect individual equipment may respond to variations in system variables and thus affect the system performance.

The most large-power systems install equipment to allow operations personnel to monitor and operate the system in a reliable manner. Some of such major types of failures are:

- Generation-unit failures
- Transmission-line outages.

Effects on Power System Due to Generation Unit Failures
When a generator experiences a failure/outage, there will be a great impact on performance of other generators and transmission lines of the PS.

Due to the imbalance between total load plus losses and generation, a drop in frequency is resulted. This must be restored back to its nominal value (50 Hz or 60 Hz). This must be made up either by the balance set of generators or by shedding a sufficient amount of load from the system. The proportion of the lost power made up by each generator is strictly determined by its governor droop characteristic, (**Appendix-C**); [6, 13]. Further, due to generation outage, much of the made up power will come from tie lines and other transmission lines. This can make line flow limit or bus voltage limit violations [13].

Effect on PS Due to Transmission Line Outages
When a transmission line or transformer fails and is disconnected, the flow on that line goes to zero and all flows nearby will be affected. The result can be a line flow limit or bus voltage limit violation [13].

The reactive losses in the transmission system (an equivalent π-model is shown in Fig. 2.2) have a big effect on the voltage at the buses. Reactive losses may be due to two reasons:

Fig. 2.2 π-model of a transmission-line

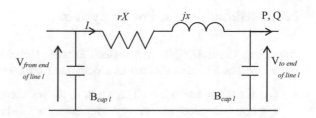

- The MVAR consumed by the line
- Transformer inductive reactance

$$\text{Reactive loss, (i)} = \sum_{all\ lines\ l} I_l^2 x_l \qquad (2.1)$$

Since,

Reactive power consumed by the transmission line $\propto$ square of the line current, when the transmission lines become heavily loaded this term goes up and more reactive power must be supplied from some other resource.

Due to capacitive charging of the transmission line, reactive power is injected back into the PS.

$$\text{Reactive loss, (ii)} = -\sum_{all\ lines\ l} \left(V_{from\ end\ of\ line\ l}^2 B_{cap\ l} + V_{to\ end\ of\ line\ l}^2 B_{cap\ l} \right) \qquad (2.2)$$

There can be fixed capacitors injecting reactive power into buses.

$$\text{Reactive loss, (iii)} = -\sum_{all\ lines\ l} \left(V_l^2 B_{fixed\ cap\ at\ bus\ l} \right) \qquad (2.3)$$

The total reactive loss = Reactive loss, (2.1) + Reactive loss, (2.2) + Reactive loss, (2.3)

$$\text{Reactive power loss} = -\sum_{all\ lines\ l} I_l^2 x_l - \sum_{all\ lines\ l} \left(V_{from\ end\ of\ line\ l}^2 B_{cap\ l} + V_{to\ end\ of\ line\ l}^2 B_{cap\ l} \right)$$
$$- \sum_{all\ lines\ l} \left(V_l^2 B_{fixed\ cap\ at\ bus\ l} \right)$$

$$\text{Reactive power loss} = \sum_{all\ lines\ l} I_l^2 r_l$$

2.5   Reliability of a Power System

According to NERC (North American Electric Reliability Council), reliability of power system has been defined as a combined process of,

- Transmission adequacy—The ability of the electric system to supply the aggregate demand and energy requirements of their customers at all times, taking into account scheduled and reasonably expected unscheduled outages of system elements/components.
- Transmission security—The ability of the bulk electric system to withstand sudden disturbances such as electric short circuits, unanticipated loss of system components or switching operations [9].

2.6   Quality of a Power System

Fundamental requirements concerning the quality of (power) generation equilibrium operation are:

- Network frequency should be at its "nominal" value (the choice of the nominal value is a technical and economic compromise among design and operating characteristics of main components, with specific regard to generators, transformers, lines, and motors);
- Voltage magnitudes (positive sequence) should match their nominal values, within a range, e.g., of ±5% or ±10% at each network bus-bar, particularly at some given load bus-bars [14].

2.6.1   Addressing Instability Situations Due to Perturbations in the Power System

Facing the effects of perturbations, especially of those lasting longer, and maintaining the system at satisfactory steady-state conditions can be done with two fundamental controls. They are:

- f/P control (frequency and active power control)—acts on control valves of prime movers, to regulate frequency and dispatch active power generated by each plant. Frequency regulation is the modulation of driving powers which must match, at steady-state conditions, the total active load (apart from some deviations due to mechanical and electrical losses, or contributions from non-mechanical energy sources). After a perturbation, the task of frequency regulation is not only to make net driving powers and generated active powers coincide but, moreover, to return frequency to the desired value. Therefore, even the regulation itself must cause transient unbalances between the powers until the frequency error returns to zero.

- *V/Q* control (voltages and reactive power control)—acts on the excitation circuit of synchronous machines and on adjustable devices (e.g., reactors, capacitors, static compensators, under-load tap-changing transformers), to achieve acceptable voltage profiles with adequate power flows in the network.

f/P and *V/Q* controls are different from each other with regard to the power system stability concern.

- Regulated frequency is common to the whole system and can be affected by all the driving powers. Therefore, the *f/P* control must be considered with respect to the whole system, as the result of different contributions (to be suitably shared between generating plants). In other words, the *f/P* control must present a "hierarchical" structure (as shown in Figs. 2.1 and 2.6) in which local controls (also named "primary" controls) on each turbine are coordinated through a control at the system level (named "secondary" control).
- Regulated voltages are instead dissimilar from each other (as they are related to different network points), and each control predominantly acts on voltages of the nearest nodes. Consequently, the *V/Q* control problem can be divided into more primary control problems (of the local type), which may be coordinated by a secondary control (at the system level) or simply coordinated at the (unit) scheduling stage [14].

2.6.2 Classification of Power System Dynamics

Dynamic relations among variables that characterize a generic model of a generator system can be presented as in the block diagram of Fig. 2.3.

Subsystem (a): Predominantly a mechanical type, consists of generating unit rotating parts (specifically, inertias) and supply systems (thermal, hydraulic, etc.).

Subsystem (b): Predominantly an electrical type, consists of the remaining parts, i.e., generator electrical circuits, transmission, and distribution systems, and users (and possible energy sources of the non-mechanical type), with the latter possibly assimilated with electrical equivalent circuits. Subsystem (b) includes mechanical rotating parts of synchronous compensators and electro-mechanical loads. The mechanical parts of synchronous compensators and of synchronous motors—the latter including their loads—can be considered, if worthy, in subsystem (a) without any particular difficulties [14].

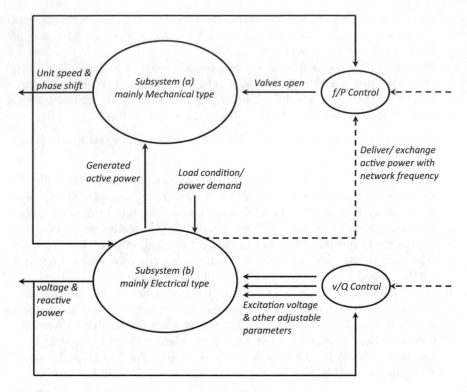

Fig. 2.3 A generic model of a generator system [14]

As demonstrated by the Fig. 2.3, the input variables to the system are essentially,

- openings of prime mover valves, which "enter" into subsystem (a), affecting driving powers (at given operating conditions of the supply systems, e.g., set points of the boiler controls, water stored in reservoirs);
- excitation voltages of synchronous machines, which "enter" into subsystem (b), affecting the amplitude of emfs applied to the three-phase electrical system;
- different parameters that can be adjusted for control purposes (specifically, for the V/Q control): capacitances and inductances of reactive components (of the static type), transformer ratios of under-load tap-changing transformers, etc.;
- load conditions dictated by users, which are further inputs for the subsystem (b), in terms of equivalent resistances (and inductances) or in terms of absorbed mechanical powers, etc.

With reference to the Fig. 2.3, the *f/P* control is achieved through acting on valves' opening, while the *V/Q* control is achieved through acting on excitation voltages and the adjustable parameters mentioned above. The load conditions instead constitute "disturbance" inputs for both types of control.

Subsystems (a) and (b) interact with each other, specifically through:

- generated active powers;
- electrical speeds of generating units (or, more generally, of synchronous machines) and (electrical) shifts between their rotors.

With reference to [14], typical time intervals for analysis and control of the most important dynamic phenomena are shown in Fig. 2.4. Regarding response times, subsystem (a) generally presents much slower "dynamics" than subsystem (b) (except with torsional phenomena on turbine-generator shafts), primarily because of:

- the effects of rotor inertias,
- limits on driving power rate of change,
- delay times by which (because of the dynamic characteristics of supply systems) driving powers match opening variations of the valves.

With the help of above facts various simplifications can be done that are useful in,

- identifying the most significant and characterizing factors of phenomena,
- performing dynamic analyses with reasonable approximation, and
- selecting the criteria and implementing on the significant variables, on which the real-time system operation (control, protection, supervision, etc.) should be based.

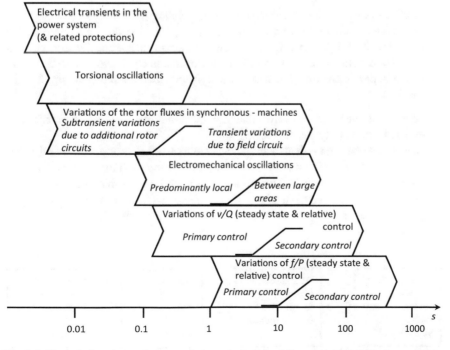

Fig. 2.4 Typical time intervals for analysis and control of the most important power system dynamic phenomena [14]

Accordingly, dynamic phenomena can be categorized as below [14].

• *Predominantly* Mechanical phenomena	− caused by perturbations in subsystem (a) and in *f/P* control, − slow enough to allow rough estimates on the transient response of subsystem (b), up to the adoption of a purely "static" model (an example is the case of phenomena related to frequency regulation)
• *Predominantly* Electrical phenomena	− caused by perturbations in subsystem (b) and in *V/Q* control, − fast enough that machine speeds can be assumed constant (for instance, the initial part of voltage and current transients following a sudden perturbation in the network) or which are such to produce negligible variations in active powers, again without involving the response of subsystem (a) (for instance, phenomena related to voltage regulation, in case of almost purely reactive load)
• *Strictly* Electromechanical-phenomena	− Caused by interaction between subsystems (a) and (b) − acceptable to simplify the dynamic models of components according to the frequencies of the most important electromechanical oscillations (e.g., oscillation of the rotating masses of the generators and motors that occur following a disturbance, operation of the protection system [6])

2.7   Process for Generation-Load Balance

An electrical power system consists of many generating units and many loads while its total power demand varies continuously throughout the day in a more or less anticipated manner. It is very important for the utilities to ensure that the power system can be catered with sufficient generation whenever it is on demand. There are four main time frames in ensuring that they can supply their loads as shown in Fig. 2.5, [9].
They are:

- Long term planning—ensures that the most optimal generation portfolio is invested into supply the forecasted load
- Operations planning—deals with changes in transmission or generation that will need to take place for maintenance purposes in the coming months. The unit function that deals with the optimum selection of the units that need to go online to supply the load may fall in this time frame as well, depending on their type, generating units need different preparation time for going live spanning from months to days.

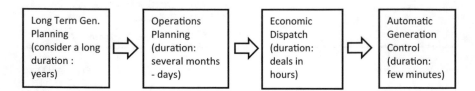

Fig. 2.5   Generation load balance in different time horizons [9]

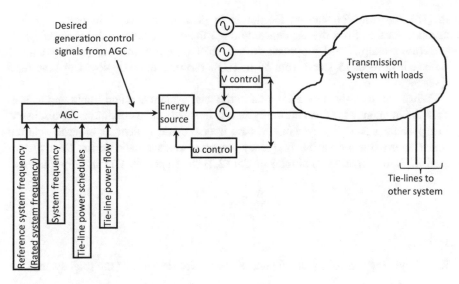

Fig. 2.6 Power system automatic generation control [9]

- Economic dispatch—deals with the selection of the most economic units to supply the load in the next few hours.
- Automatic Generation Control, AGC—balance generation and load on a minute-to-minute basis when operators do not have sufficient time to control generators. With reference to AGC, Load Frequency Control is demonstrated in Fig. 2.6, considering one generator in the system.

According to [15], Load frequency control is described in the "UCTE Operation Handbook" as "the continuous balance between supply and demand that must be maintained for reliability and economic operational reasons." The system frequency, which should not vary significantly from its set point of 50 Hz, is an indication of the quality of "balance". The load frequency control can be identified with five control levels. They are:

- Primary control
- Secondary control
- Tertiary control
- Time control
- Measures for emergency conditions.

2.7.1   Primary Control (Is by Governors)

The action of turbine governors due to frequency changes when reference values of regulators are kept constant is referred to as primary frequency control. According to [15], the time for starting the action of primary control is in practice a few seconds starting from the incident (although there is no intentional time delay for governor pickup), the deployment time for 50% or less of the total primary control reserve is at most 15 s and from 50 to 100% the maximum deployment time rises linearly to 30 s.

When the total generation is equal to the total system demand (including losses) then the frequency is constant, the system is in steady state condition. As discussed in Appendices A–C [6], system loads are frequency dependent. In order to obtain a linear approximation of the frequency response characteristic of the total system load, a similar expression similar to Eq. (2.2) (in Appendix-C) can be written as,

$$\frac{\Delta P_L}{P_L} = K_L \frac{\Delta f}{f_n} \tag{2.4}$$

where,

K_L frequency sensitivity coefficient of the power demand of the total system

From Eq. (2.2) (in Appendix-C),

$$\frac{\Delta P_T}{P_L} = -K_T \frac{\Delta f}{f_n} \tag{2.5}$$

Tests conducted on actual systems indicate that the generation response characteristic is much more frequency dependent than the demand response characteristic.

Typically,

$$K_L = \text{between } 0.5 \text{ and } 3$$
$$K_T \approx 20(\rho = 0.05).$$

In Eqs. (2.2) and (2.4) the coefficients K_T and K_L have opposite signs so that an increase in frequency corresponds to a drop in generation and an increase in electrical load.

In the (P, f) plane the intersection of the generation and the load characteristic, Eqs. (2.2) and (2.4), defines the system equilibrium point.

A change in the total power demand ΔP_L corresponds to a shift of the load characteristic in the way shown in Fig. 2.7, so that the equilibrium point is moved from position 1 to position 2. The increase in the system load is compensated in two ways:

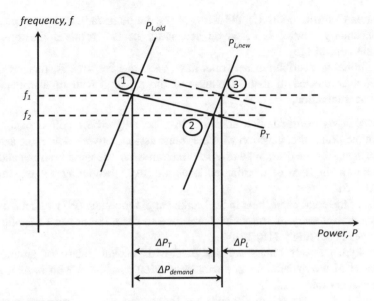

Fig. 2.7 Equilibrium points for an increase in the power demand [6]

1st   by the turbines increasing the generation by ΔP_T.
2nd   by the system loads reducing the demand by ΔP_L from that required at
      position 3 to that required at position 2

$$\Delta P_{demand} = \Delta P_T - \Delta P_L = -(K_T - K_L)P_L \frac{\Delta f}{f_n} = K_f P_L \frac{\Delta f}{f_n} \qquad (2.6)$$

where,

K_f stiffness of a given area or power system.

A reduction of the demand by ΔP_L is due to the frequency sensitivity of demand. An increase of generation by ΔP_T is due to turbine governors. **The action of turbine governors due to frequency changes when reference values of regulators are kept constant is referred to as primary frequency control.**

2.7.2 Secondary Control (Is by Automatic Generation Controls)

This maintains a balance between generation and consumption (demand) of the power system as well as the system frequency, without disturbing the primary control that is operated in the corresponding power system in parallel, but by a margin of seconds. Secondary control makes use of a centralized Automatic

Generation Control, modifying the active power set points/adjustments of generator sets. Secondary control is based on secondary control reserves that are under automatic control [15].

Traditionally, distribution networks have been passive, that is, there was little generation connected to them. Because of the rapid growth in distributed and renewable generation,

- Power flows in distribution networks may no longer be unidirectional, that is from the point of connection with the transmission network down to customers. In many cases the flows may reverse direction when the wind is strong and wind generation high, with distribution networks even becoming net exporters of power [6].
- Most of the solar plants here in Sri Lanka are photovoltaic (PV) with an inverter. Since the intensity of sun varies from time to time, the power generation also changes accordingly [16, 17].
- Mini-hydro Power Plants are designed with a plant factor of around 40% because of the uncertainty in power generation associated with it, as it totally depends on rain fall [5].
- Hydro Power generation depends on the stored energy (reserved water) in corresponding reservoirs [5, 6, 16, 18].

Therefore, the direction of the current flow in transmission mesh network varies from time to time [19].

Further, with the state-of-the-art wind forecasting methods, the hour ahead forecast errors for a single wind power plant are still around 10–15% with respect to its actual outputs [20].

That situation has created many technical problems with respect to settings of protection systems, voltage drops, congestion-management etc.

Hence, the main Duties performed by Automatic Generation Control are:

- maintain frequency at the scheduled value (frequency control);
- maintain the net power interchanges with neighboring control areas at their scheduled values (tie-line control);
- Maintain power allocation among the units in accordance with area dispatching needs (energy market, security or emergency).

In certain systems, one or two of the above objectives may be handled by the AGC. For example, tie-line power control is only used where a number of separate power systems are interconnected and operate under mutually beneficial contractual agreements [6].

This is clearly demonstrated in Fig. 2.6, where AGC measures,

- Actual system frequency
- Interchange flows

From which it calculates,

- the frequency and
- interchange flow deviations,

by using the reference frequency and scheduled interchange values. The frequency and interchange deviation are then used to balance load and generation on a minute-to-minute basis.

Different Automatic Generation Control applications such as:

- Governor Control System
- Interconnected Operation

are implemented [6, 9].

Governor Control System

The slope of the curve shown in Fig. 2.8 is known as the governor droop. That is:

$$\text{Governor droop} = \frac{(f_0 - f_1)}{(G_0 - G_1)}$$

In general,

$$\text{Governor droop} = -\frac{\Delta f}{\Delta G} \, \text{Hz/MW} \tag{2.7}$$

$$\text{p.u. Governor droop } \% = -\frac{\Delta f / f_0}{\Delta G / G_R} \tag{2.8}$$

where,

f_0 rated frequency
Δf system frequency change
G_R rated generation capacity
ΔG system generation change

Fig. 2.8 A typical speed power characteristic of a governor system

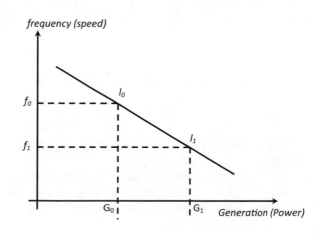

∴ when the droop increases, the generation response is less sensitive to a frequency change. As the system frequency is a constant all over a given system, from the Eqs. (2.8) and (2.9) it can be understood that, for a Δf change in system frequency, response from different generating units are:

$$\frac{\Delta G_i}{G_R} = \frac{-\Delta f/f_0}{GD_i} \Rightarrow \Delta G_i = -(\Delta f)\frac{G_R/f_0}{GD_i} \tag{2.9}$$

where,

ΔG_i generation change
GD_i governor—droop of the ith generator

As a result of the frequency change, total generation change in an n generator system is:

$$\text{Total generation change in MW} = \sum_{i=1}^{n} \Delta G_i = \sum_{i=1}^{n} -(\Delta f)\frac{G_{Ri}/f_0}{GD_i} \tag{2.10}$$

As a result of the frequency change the system load also changes due to load sensitivity to frequency.

$$\Delta L_i = D\Delta f \tag{2.11}$$

where,

ΔL_i Load change (MW)
Δf frequency change (Hz)
D Load frequency variation factor (MW/Hz)

From Eqs. (2.10) and (2.11),

$$\text{Total generation change in MW} = \sum_{i=1}^{n} \Delta G_i - \Delta L_i = \sum_{i=1}^{n} -(\Delta f)\frac{G_{Ri}/f_0}{GD_i} - D\Delta f \tag{2.12}$$

$$\text{Total generation change in MW} = \sum_{i=1}^{n} \Delta G_i - \Delta L_i = -(\Delta f)\left(\sum_{i=1}^{n} \frac{G_{Ri}/f_0}{GD_i} + D\right) \tag{2.13}$$

Hence due to a frequency change of Δf, there will be an adjustment in both load and generation as given in Eq. 2.13.

As shown in Fig. 2.9, changes in the settings $P_{ref(1)}$, $P_{ref(2)}$ and $P_{ref(3)}$ enforce a corresponding shift of the characteristic to the positions $P_{m(1)}$, $P_{m(2)}$ and $P_{m(3)}$. A turbine cannot be forced to exceed its maximum power rating P_{MAX} with change

Fig. 2.9 Turbine speed–droop characteristics for various settings of P_{ref}

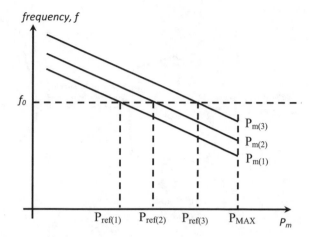

of settings. Further Changing in settings P_{ref} of individual governors will move upwards the overall generation characteristic of the system. Eventually this will lead to the restoration of the rated frequency but now at the required increased value of power demand. Such **control action on the governing systems of individual turbines is referred to as secondary control.**

Interconnected Operations

Power systems interconnections are put in place for different systems to be able to

- Perform exchange of electricity and enjoy the economic benefits of diversity in generation and load
- Provide support under contingencies

Since load and generation in each system change instantaneously, it is important to have proper controls on interties. These controls ensure that the undesirable tie-line flows do not show up as the systems try to mitigate frequency deviations. In other words, each system provides its share of frequency correction without impacting another system's generation load balance inadvertently [6, 9].

In interconnected power systems, AGC is implemented in such a way that each area, or subsystem, has its own central regulator. As shown in Fig. 2.10, the power system is in equilibrium if, for each subsystem satisfy the condition,

$$P_T - (P_L + P_{tie}) = 0 \tag{2.14}$$

2.7.3   Tertiary Control

This action restores secondary control reserve by rescheduling generation and is put into action by the responsible undertakings. The task of tertiary control depends on

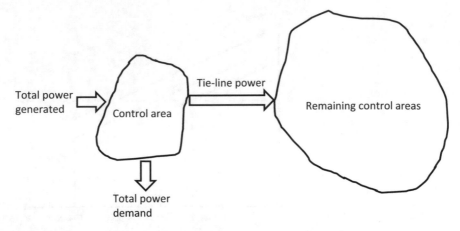

Fig. 2.10 Power balance of a control area [6]

the organizational structure of a given power system and the role that power plants play in this structure. Tertiary control is an additional frequency control procedure to primary and secondary frequency control. This is slower than, primary and secondary frequency control [6, 15].

Under the vertically integrated industry structure (see Fig. 2.1), the system operator sets the operating points of individual power plants based on the economic dispatch, or more generally Optimal Power Flow, which minimizes the overall cost of operating plants subject to network constraints. Hence tertiary control sets the reference values of power in individual generating units to the values calculated by optimal dispatch in such a way that the overall demand is satisfied together with the schedule of power interchanges.

In many parts of the world, electricity supply systems have been liberalized. Gradually, private owned power plants are added to the power system of Sri Lanka too [21, 22]. These privately owned power plants are not directly controlled by the system operator. Instead, according to their mutual agreements with the government or utility, the generation of electricity is practiced by them. Hence the economic dispatch is executed through an energy market.

The main task of the system operator is then to adjust the contracts to make sure that the network constraints are satisfied and to supply the required amount of primary and secondary reserve from individual power plants. In such a market structure the task of tertiary control is to adjust, manually or automatically, the set points of individual turbine governors in order to ensure,

- Adequate spinning reserve in the units participating in primary control.
- Optimal dispatch of units participating in secondary control.
- Restoration of the bandwidth of secondary control in a given cycle.

(The sum of regulation ranges, up and down, of all the generating units active in secondary control is referred to as the **bandwidth of secondary control**. The positive value of the band-width, that is, from the maximum to the actual operating point, forms the reserve of secondary control [6].)

Tertiary control is supervisory with respect to the secondary control that corrects the loading of individual units within an area. Tertiary control is executed through,

- Automatic change of the reference value of the generated power in individual units.
- Automatic or manual connection or disconnection of units that are on the reserve of the tertiary control [6].

Usually control areas are grouped in large interconnected systems with the central regulator of one area (usually the largest) regulating power interchanges in the given area with respect to other areas. In such a structure the central controller of each area regulates its own power interchanges while the central controller of the main area additionally regulates power interchanges of the whole group.

2.7.4 Time Control

This action corrects global time deviations of the synchronous time in the long term as a joint action of all undertakings.

It is considered that each utility's load is composed of,

- its native load and
- the scheduled transactions with other utilities.

Two numbers of system measurements are used to reflect the degree of generation-load imbalance. They are [9],

- the system frequency—which is constant in the whole system and reflects whether the native load is balanced. By removing the frequency error, the AGC control ensures that the generation and load are balanced out.
- the total interchange obligation the utility has to other utilities—if a number of interconnected utilities all attempt to remove the system frequency deviations, they will end up balancing load and generation for the whole interconnected system, except that the final outcome may not fulfill their interchange obligations to each other.

Area Control Error (ACE)

As discussed above AGC controller should minimize both the frequency and interchange deviations as its objective function in its control design. Hence, Area

Control Error (ACE) is defined for a system. Within each control area, continuously, this ACE should be controlled to zero.

$$ACE = -10 \cdot \beta_f \cdot (f_{Actual} - f_{Desired}) + (T_{Actual} - T_{Scheduled}) \qquad (2.15)$$

where,

$f_{Desired}$ desired frequency (e.g. 50 Hz)
f_{Actual} actual frequency in Hz
T_{Actual} actual interchange schedule or tie-line flow with a positive sign for export, in MW
$T_{Scheduled}$ scheduled interchange or tie-line flow with a positive sign for export, in MW
β_f area bias with a negative value, MW per 0.1 Hz

AGC achieves its objective by minimizing or bounding ACE. It is important to note that ACE only considers the status of a snapshot of the system disregarding the frequency deviations in the previous time periods and the accumulated frequency and interchange flow mismatches. This creates other issues associated with accumulated frequency and interchange deviations as shown in Fig. 2.7. These frequency deviation and interchange deviation can be reduced to zero by AGC.

Time Error
All electric clocks operates on the main system, are based on system frequency. As shown in Fig. 2.11, the error with time can be considered as a function of accumulated frequency deviation, which is reflected by the area under the frequency deviation curve.

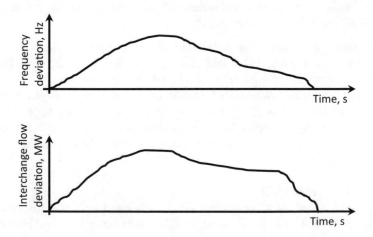

Fig. 2.11 Frequency and interchange flow deviations for one area [1]

To ensure that the time shown by electric clocks is correct, system operators should correct the accumulated time error by scheduled actions. The time error resulting from frequency deviation can be calculated by,

$$Time\ error = \int \left(\frac{f_{Actual} - f_{Desired}}{f_{Desired}}\right).dt \quad (2.16)$$

where,

$f_{Desired}$ desired frequency in Hz
f_{Actual} actual frequency in Hz

The time error adjustment is not accomplished by AGC. It is sometimes added to the ACE Eq. 2.16 for the sake of completeness, as given in Eq. 2.17:

$$ACE = -10.\beta_f.(f_{Actual} - f_{Desired}) + (T_{Actual} - T_{Scheduled}) + \beta_t.(Time\ error) \quad (2.17)$$

β_t the time error bias in MW/s

In any given situation, the responses of only a limited amount of equipment may be significant. Therefore, using assumptions, usually it is made to simplify the problem and to focus on factors influencing the specific type of stability problem [11]. Further, in analyzing more complex cases where simplifications may not seem acceptable, computer simulations can become necessary [14].

2.8 Under-Frequency Load Shedding

AGC as a "secondary control" has been used by power systems for several decades for bringing up the system frequency to its nominal value, with its actions usually slower than the "primary control" which is done by turbine governors. In any event, the first seconds of frequency dip and recovery after a major generator trip is essentially be accomplished by governor control. When the power system's self-regulation is insufficient to establish a stable state, the system frequency will continue to drop until it is arrested by automatic under-frequency load shedding (UFLS) to re-establish the load-generation balance within the time constraints necessary to avoid system collapse.

This can be considered as one of the most possible contingencies that may lead the Power System unstable. Events that can be identified as critical signals are:

- Open Circuit of generator feeder or grid transformer Circuit Breaker;
- OC of a bus-coupler feeder or tie line Circuit Breaker;
- Protection lock-out function operation of a critical Circuit Breaker;
  (In LS functionality, grid or generator or network Circuit Breakers are referred to as critical Circuit Breakers.);

- Hidden failures in protection systems [3, 12, 23, 24].

The problem of optimal LS has been extensively investigated and many publications on the utility implementation are presented in the literature over the past, [3, 10, 12, 23, 25, 26]. Selection of a suitable LS strategy depends on the application scenario.

- In large scale and wide area Power Systems, typically the adopted methods are based on voltage measurement. This helps in determining the perturbation location so that the area affected by the power deficiency can be addressed by implementing a Load Shedding action confined to that particular area.
- In local power systems, detection of the location of the contingency is trivial. For such situations, LS actions are implemented mainly based on frequency and its derivatives [3, 10, 27].

As demonstrated in Fig. 2.12, LSS acts whenever it identifies a situation of danger for the PS. The most initiative method of checking the level of danger is measuring the average frequency of the grid: when the frequency falls below a certain threshold it is possible to obtain an indication on the risk for the system and consequently to shed a certain amount of load. Although this approach is effective in preventing inadvertent LS in response to small disturbances with relatively longer time delays and low frequency thresholds, it is not capable of distinguishing the difference between normal oscillations and large disturbances in the power system. Thus, this approach is prone to shed lesser loads at large disturbances.

During a load and generation imbalance situation that occurred due to a generation deficiency, the amount of over-load is not known. Therefore the load is shed in blocks until the frequency stabilizes. The three main categories of LS methodologies are: Traditional, semi-adaptive and adaptive.

- Traditional LS scheme is mostly implemented because of its simplicity and less requirement of sophisticated relays. It sheds a certain amount of load when the system frequency falls below a threshold. If this load drop is sufficient, the frequency will stabilize or increase. This process continues until the overload relays get operated. The value of the threshold and the relative amount of load to

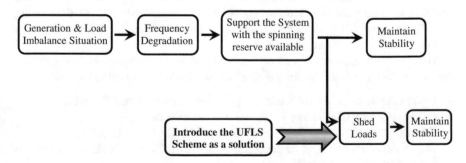

Fig. 2.12 Power system responses due to a load-generation imbalance situation

be shed are decided off line based on experience and simulation. Although this approach is effective in preventing inadvertent LS in response to small disturbances with relatively longer time delay and lower frequency threshold, it is not able to distinguish between normal oscillations and large disturbances of the power system. Thus, this approach is prone to shed lesser loads at large disturbances [8, 28].

- The semi-adaptive LS scheme uses the frequency decline rate as a measure of the generation shortage. In this scheme, the rate of change of frequency thresholds and the size of load blocks to be shed at different thresholds are decided off-line on the basis of simulation and experience [8, 28].

- Adaptive LS scheme is the one that can prevent black-outs through controlled disintegration of a power system into a number of islands together with generation tripping and/or LS [8, 28, 29]. In [29], a linear System Frequency Response (SFR) model is developed which is based on the frequency derivative of the Power System. According to [28, 29], from the reduced order SFR model it is possible to obtain a relation between the initial value of the ROCOF and the size of the disturbance P_{step}, which caused the frequency decline. This relation is:

$$\frac{df}{dt}\Big|_{t=0} = \frac{P_{step}}{2H}$$

where,

F expressed in per unit, on base of the nominal system frequency (50 Hz)

P_{step} in per unit on the total MVA of the whole system

The initial value of the ROCOF is proportional, through the inertia constant H, to the size of the disturbance. Thus, assuming that the inertia of the system is known, the measure of the initial ROCOF is—through H—a backward estimate of the disturbance and consequently an adequate countermeasure in terms of load-shedding can be operated. A drawback of this method is that, if generators or large synchronous motors are disconnected during the disturbance, the inertia of the system should be accordingly adapted. For large systems, this can be overcome by the consideration that only a small percentage of the total inertia has been lost [28, 29]. For small systems such as Power system of Sri Lanka, this may generate an under estimation of the actual perturbation.

Different methodologies have been introduced to implement LS actions based on frequency [8, 12, 23, 25, 26, 30, 31]. With reference to them, it is possible to understand an intelligent and adaptive, control and protection system for wide-area disturbance is needed, to make possible full utilization of the power network, which will be less vulnerable to a major disturbance.

Adaptive settings of frequency and frequency derivative (ROCOF) relays, based on actual system conditions, may enable more effective and reliable implementation of LSSs [10, 27, 32]. A major component of adaptive protection systems is their ability to adapt to changing system conditions. Thus, relays which are going to participate in the process of control and protection must necessarily be adaptive. In precise, this must be a relay system that allows communication with the outside word. These communication links must be secure, and the possibility of their failure must be considered in designing the adaptive relays [27].

The problem of optimal load shedding has been extensively investigated. As the power systems are dynamic and difficult to model in advance, control schemes should be capable of adjusting their decision criteria/parameters adaptively and independently. In [12], the 'Reinforcement learning method' has been introduced to provide suitable basis for the adaptation and the 'Temporal difference learning method' has been implemented for the Load Shedding Scheme to provide the reinforcement function. A methodology to develop a reliable load shedding scheme for power systems with high variability and uncertainty, under any abnormal condition, has been introduced, to prevent black-outs while maintaining its stability in [23]. The technique proposes the sequence and conditions of applications of different load shedding schemes and islanding strategies. Another load shedding scheme is proposed in [33] which Develops an algorithm for the selection of Load circuits to shed in periods of reduced generation, that may occur owing to industrial action or if the installed generation capacity is insufficient to meet demand. In [32], an adaptive centralized under-frequency load shedding scheme is described. The frequencies measured by Phasor Measurement Unit are used for calculating the rate of change of frequency as well as the magnitude of the disturbance in the power system. This method can estimate the magnitude of overload occurring from different disturbances and accordingly to determine the necessary amount of load to be shed as well as the size and frequency setting of each shedding step. Since balancing of frequency of an islanded system is still an issue to be solved, especially when the demand exceeds the generation in the power island, a strategy to shed an optimal number of loads in the island to stabilize the frequency is presented in [25]. In [27], Miroslav Begovic et al. explore special protection schemes and new technologies for advanced, wide-area protection. There it has been high-lighted that there is a great potential for advanced wide-area protection and control systems, based on powerful, flexible and reliable system protection terminals, high speed, communication, and GPS synchronization in conjunction with careful and skilled engineering by power system analysts and protection engineers in cooperation. Some of the basic principles which should be considered in the application of a load shedding and load restoration program are presented in [34]. The philosophy which led to the frequency actuated load shedding and load restoration program being implemented on the American Electric Power System are also discussed. In [26], a new approach to adaptive UFLS based on frequency and rate of change of frequency is presented, which are estimated by non-recursive Newton type algorithm. A load shedding scheme for the power system of Sri Lanka considering a coal power plant with 300 MW generation capacity is suggested in [30]. Draw backs of the then load shedding scheme are also presented. In [28], several load-shedding schemes for under-frequency operation are examined. Both traditional schemes, based only on frequency thresholds, and adaptive schemes, based on frequency and on its rate of change, are considered. An IEEE test system for reliability analysis is used to compare the behavior of the proposed schemes when selecting different thresholds and percentages of load to be disconnected. Results are reported in detail; considerations on possible advantages and drawbacks are also related to the framework provided by the electricity market. M. Giroletti et al., propose a new hybrid Load Shedding method, which combines the most significant features of frequency-based and power-based LS approaches in their publication [10]. A solution technique is described for designing a load shedding scheme using under frequency relays to limit the effect of system over loads and a description of each necessary design decision is also presented in [31]. In their literature, the improvements done in calculating the tripping frequency of a given load shedding step, using the clearing frequency of the previous step and the use of a per-unit relative efficiency ratio are also presented. This method guarantees co-ordination between load-shedding steps. In [25], a technique to develop a frequency dependent auto load-shedding and islanding scheme to bring a power system to a stable state and also to prevent blackouts under any abnormal condition is described. The technique incorporates the sequence and conditions of the application of different load shedding schemes and islanding strategies. The technique is developed based on the international current practices. It uses the magnitude and the falling rate of change of frequency in an abnormal condition to determine the relay settings offline. The paper proposes to implement the technique using only Frequency Sensitive and Frequency Droop relays. For developing countries located in the South Asian region such as Sri Lanka and India (which have poor power systems) may experience equal types of power system instability situations which lead for catastrophic events as in Bangladesh. So the above technique would be a good solution to eliminate power system black-outs while maintaining the stability of the power system. But,

- the traditional LS scheme incorporated with the above technique suggests some time-delay and these time-delays vary from one step to another; hence deciding a correct time delay for corresponding LS stage may be a practically difficult task,
- even though it suggests to disintegrate the power system and to operate in the islanding mode if the frequency decline rate exceeds a particular threshold, power system instability situations can be occurred during the disintegration of the system [12, 23].
- Further when a disturbance occurs, it takes some time to reach particular frequency degradation (df/dt). So during that period, a load shedding action can be taken place (based on traditional load shedding), before this df/dt action gets activated. This may lead for excessive load shedding.

One can look at the system operation as explained in Fig. 2.13, which has been introduced by Fink and Carlsen. In this model, conventional protection and control is likely to be effective in the 'alert' and 'emergency' states where the load capacity

Fig. 2.13 Fink and Carlsen
diagram [35]

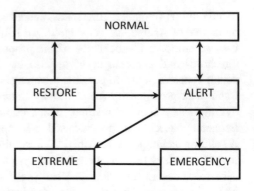

and generating capacity remain matched. In the 'extreme' state, they are no longer matched and system integrity protection schemes are required [35].

Hence, as a solution for the above requirements, the proposed methodology is presented. It is a combination of all three basic-categories of load shedding schemes mentioned above. Thereby it is aimed to provide a quality and reliable power supply for the consumer while supporting the economy of the country.

Chapter 3
Modelling the Power System

To study the stability and to observe how the power system behaves, during normal and abnormal conditions, availability of a simulation model of the considering Power System is very important. To ascertain that the simulation model's performance is identical or approximately equal to the corresponding real time power system's behavior, it has been decided to consider the power system of Sri Lanka of which required data could be collected from the Ceylon Electricity Board (CEB) on request. Therefore a simulation model of the Power System of Sri Lanka (Transmission network—132 and 220 kV) was designed using the software PSCAD/EMTDC.

PSCAD is a power-system simulator for the design and verification of power quality studies, power electronic design, distributed generation, and transmission planning. It was developed by the Manitoba HVDC Research Center and has been in use since 1975. PSCAD is a graphical front end to EMTDC for creating models and analyzing results. In PSCAD, one combines blocks to form a power network. These blocks are actually FORTRAN code, which call for an EMTDC code library to combine them into executable files. Running these files runs the simulations, and the results can be picked up by PSCAD on the run [36–39].

PSCAD is suitable for modelling the power system of Sri Lanka for several reasons. PSCAD has fully developed models of various devices used in the PS of Sri Lanka.

- The library includes models of synchronous machines, turbines, governors, transformers, relays, breakers, cables, and transmission lines. Saturation, magnetizing, and leakage inductances can be disabled or enabled in the rotating machine models. PSCAD also offers tools to simulate various faults on the power system. By providing actual parameters/data where necessary, it is possible to develop models of different power system components.
- The systems of the PSCAD library may consist of electrical and control-type components, which may be interconnected to allow for an all-inclusive simulation study. The control systems modeling function section of the PSCAD library provides a complete set of basic linear and nonlinear control

© Springer Nature Singapore Pte Ltd. 2018
T. Bambaravanage et al., *Modeling, Simulation, and Control of a Medium-Scale Power System*, Power Systems, https://doi.org/10.1007/978-981-10-4910-1_3

components. These components can be combined into larger, more elaborate systems. Outputs from control components can be used to control voltage and current sources, switching signals etc. Control components can also be used for signal analysis, and outputs may be directed to online plots or meters.

- Multiple modules can be built inside a single project, and each module can contain other modules. This provides a hierarchical modeling capability.
- PSCAD has complete models of most of the devices used in the 'simulation model of the PS of Sri Lanka'. Graphing, plotting, and exporting results are easy.

Therefore, PSCAD is suitable for this simulation study.

3.1 Power System Components

For large-scale and complex systems, the mathematical description is nonlinear and high dimensional, consisting of a large number of nonlinear equations. Therefore the analysis and computation of such a system should be started from simple local devices and be completed finally for the complex overall system. Therefore, in modeling of large-scale and complex power systems, these systems are first decomposed into independent basic components, such as:

- Synchronous Generators
- Transformers
- Transmission lines
- Under-ground cables
- Governor
- Turbines etc.

Since these components are already modeled as blocks in PSCAD, it is possible to configure them referring to actual/real time parameters of the power system. Further in configuring these mathematical models corresponding to different components, a sound knowledge of their performance and their effects on the power system must be thoroughly studied. These Models of those components are building-bricks to construct the mathematical model of the whole power systems. Even though the size and structural components of a power system vary from one to another, they all have the same basic characteristics.

3.2 Configuring Power System Components/Mathematical Modeling

For the study of some specific problems, model parameters could be time-variable and variables may not be continuous. To meet the requirement of different computing accuracy, different models could be used. A mathematical model for qualitative analysis could be simpler than that for quantitative analysis. Since computing accuracy and speed are always two conflicting factors, it is very important to consider whether a very high accuracy is required with the results, when a power system model is established [40].

There are two major issues in mathematical modelling [40]. This can be brief according to the Fig. 3.1.

As demonstrated in Fig. 3.1, Issues of mathematical modelling of a power system/power system components can be identified as:

- To describe a subject under investigation mathematically in the form of equations

  - Analytical method

    to derive those mathematical descriptions by using special knowledge and theory about the subject

  - Experimental method

    To identify those mathematical descriptions by carrying out experiments or using data obtained from its operation—the method of system identification in control theory

- To obtain parameters of the mathematical description of the subject:

  - Based on design parameters (theoretical derivation)

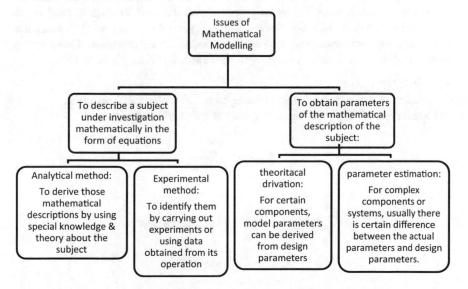

Fig. 3.1 Issues of mathematical modelling of a power system/power system components

For simple components, model parameters can be derived from design parameters according to certain physical (e.g. mechanical or electrical) principles; e.g. transmission line modeling

- Based on Parameter estimation

For complex components or systems, usually there is certain difference between the actual parameters and design parameters; e.g. generator modelling (these generator parameters can be affected by variations of the power system operating conditions, saturation and a series of complex conversion processes among mechanical, electrical, magnetic and thermal energy). This is known as 'parameter estimation' which is one of the methods in system identification.

3.2.1 Transmission Lines

Transmission lines are the main corridor of power transmission in a power system. The '*Generation and Transmission network*' as in 2011 in the power system of Sri Lanka [19] has been referred, **Appendix-D**.

Conductor Types Available in the Power System
It has been identified that different types of conductors are utilized in transmission of power [19]. All are 3-phase circuits. Some of them are single circuits while the others are double circuits. Table 3.1 gives a detailed list of conductors used in the transmission network and their corresponding per unit values of +ve sequence resistance, +ve sequence reactance and +ve sequence susceptance. These values were calculated based on the existing transmission line data given in [41].

Sample Calculation
Calculation of a p.u. value corresponding to a 220 kV transmission line:
Calculation of Z_{base} corresponding to 220 kV transmission line:

$$Z_{base} = \frac{V_{base}^2}{MVA_{base}}$$

where,

$$MVA_{base} = 100 \text{ and } V_{base} = 220\,\text{kV}$$

$$Z_{base} = \frac{220^2}{100} = 484\,\Omega$$

$$\frac{Resistance}{circuit} in\ p.u. = \frac{resistance\ of\ the\ circuit}{Z_{base}}\Omega$$

Table 3.1 Conductor types used in the transmission network of Sri Lanka

Conductor	Voltage capacity (kV)	+ve sequence resistance pu/m (Ω)	+ve sequence reactance pu/m (Ω)	+ve sequence susceptance pu/m (Ω)
lynx	132	1.016×10^{-6}	2.353×10^{-6}	4.867×10^{-7}
Oriole	132	1.0962×10^{-6}	2.5022×10^{-6}	4.8398×10^{-7}
Zebra	132	4.332×10^{-7}	2.202×10^{-6}	5.193×10^{-7}
Goat	132	5.739×10^{-7}	2.2383×10^{-6}	5.1244×10^{-7}
Tiger	132	1.4176×10^{-6}	2.4047×10^{-6}	4.7477×10^{-7}
2 × Zebra	132	2.1667×10^{-7}	1.7259×10^{-6}	6.524×10^{-7}
2 × Goat	132	2.864×10^{-7}	1.7389×10^{-6}	6.5256×10^{-7}
Zebra	220	1.5597×10^{-7}	8.5371×10^{-7}	1.3353×10^{-6}
2 × Zebra	220	7.8×10^{-8}	6.2133×10^{-7}	1.8124×10^{-6}
2 × Goat	220	1.0312×10^{-7}	6.276×10^{-7}	1.7937×10^{-7}

The transmission line from Randenigala to Rantembe has been taken as a sample for model calculations [19].

Type of circuit	Single circuit
Names of transmission lines	Rand-Rant
Steady state frequency	50.0 Hz
Length of transmission line	3.1 km
Number of conductors (per circuit)	3 nos.
Conductor type	2 × Zebra
Voltage capacity	220 kV

Referring to the data given in [41], +ve sequence resistance (R), reactance (X) and suceptance (Y) values of transmission lines were calculated. Table 3.2 gives the corresponding data used for the transmission line 'Rand-Rant' parameter calculation.

Calculation of +ve sequence resistance, +ve sequence reactance and +ve sequence susceptance:

$$+ve\ sequence\ resistance = \frac{0.00024}{3.1 \times 10^3}\ \text{pu/m} \cong 7.8 \times 10^{-8}\ \text{pu/m}$$

Table 3.2 RXY values as given in [41], for the transmission line 'Rand-Rant'

Conductor	Voltage capacity (kV)	Resistance (R)/circuit in p.u. (Ω)	Reactance (X)/circuit in p.u. (Ω)	Susceptance (Y)/circuit in p.u. (Ω)
2 × Zebra	220	0.00024	0.00196	0.00605

$$+ve\ sequence\ reactance = \frac{0.00196}{3.1 \times 10^3}\ \text{pu/km} \cong 6.2133 \times 10^{-7}\ \text{pu/m}$$

$$+ve\ sequence\ susceptance = \frac{0.00605}{3.1 \times 10^3}\ \text{pu/km} \cong 1.8124 \times 10^{-6}\ \text{pu/m}$$

Values Used with PSCAD Window
In configuring the transmission lines, the Bergeron model was used since,

- It is useful for studies to get the correct steady state impedance/admittance of the transmission line at a specified frequency, and it is not used in analysis of transient studies or harmonic behavior.
- It is a very simple and constant frequency model based on travelling waves [37].

Figure 3.2 shows the PSCAD windows where the corresponding transmission line parameters were entered in configuring the transmission line—'Rand-Rant'. Figure 3.3 demonstrates as it appears in the above transmission line has been simulated in the Sri Lanka power system simulation model.

Figure 3.3 is a small part of the large transmission network of the power system of Sri Lanka that was extracted from the simulation model done. It shows how the bus-bars of power stations randenigala and Rantembe are linked through the transmission line 'Rand-Rant.'

3.2.2 Under-Ground Cables

In certain places of the power system, it has been identified the power transmission is done through underground cables. These cables are with different current capacities [42, 43]. In configuring these under-ground cables in the PSCAD software, factors such as

- their construction criteria
- insulation materials used
- how they are laid down etc. are considered.

Cable Types Available in the Power System
With reference to [19], the types of cables used for power transmission in the power system of Sri Lanka are:

- XLPE 150
- Cu 500
- 1000 XLPE
- 800 XLPE
- Cu 350

Fig. 3.2 PSCAD windows corresponding to transmission line from Randenigala to Rantembe named as 'Rand-Rant' with it's **a** line model general data; **b** Bergeron model options; **c** manual entry of Y, Z

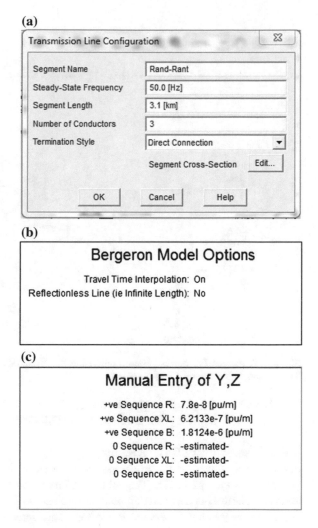

Unlike over-head transmission conductors, the under-ground cables are comprised with several layers of insulation and protective materials. With reference to [42], XLPE cable components are demonstrated in Fig. 3.4.

It has been assumed that the conducting materials in all cables are copper. Details of the cable physical parameters, resistivity of metals used and permittivity of insulating materials used are presented in Tables 3.3, 3.4, and 3.5 respectively [42, 43].

The voltage of a cable circuit is designed in accordance with the following principles: Eg: U_0/U (U_m): 130/225 (245)

where,

$U_0 = 130$ kV, phase to ground voltage

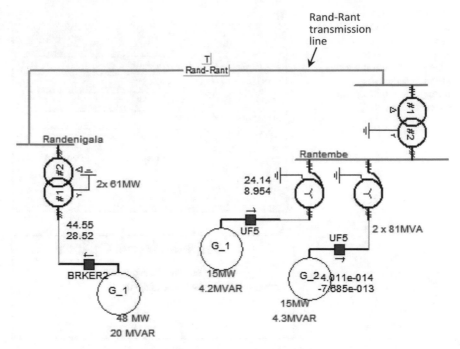

Fig. 3.3 A part of the power system comprising the 'Rand-Rant' transmission line that links Randenigala and Rantembe power-stations

U = 225 kV, rated phase to phase voltage
Um = 245 kV, highest permissible voltage of the grid

Values Used with PSCAD Window
Figure 3.5 is a part of the transmission network of the simulation model of the power system of Sri Lanka. The cable that links Colombo Fort substation (Sub-F) and the Kelanitissa power station bus-bars are shown in it. Even though the Fig. 3.4 demonstrates several layers of insulations and screen/sheath in addition to the conductor, for this simulation model it has been assumed that,

- Each cable is comprised of conductor, insulator-1, insulator-2 and sheath only.
- All three cables are placed horizontally on the same level of the ground
- Span between two cables (center to center) = 2 × cable diameter [42].
- Each cable carries a current of single phase (for three phases, 3 nos. separate cables have been used)

Accordingly, how such a cable (Keltissa-Col_F) has been configured in PSCD is demonstrated in Fig. 3.6.

Cable configuration window, Line model general data and Bergeron Model Options corresponding to 'Keltissa-Col_F' are demonstrated in Fig. 3.6a, b and c respectively. Figure 3.7 shows how the above 132 kV under-ground cable have been laid down in earth.

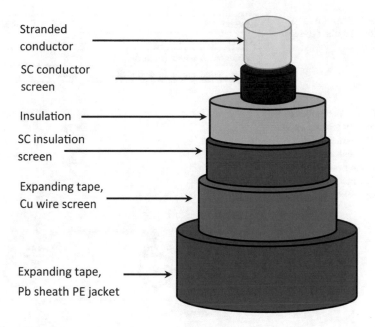

Stranded conductor

SC conductor screen

Insulation

SC insulation screen

Expanding tape, Cu wire screen

Expanding tape, Pb sheath PE jacket

Fig. 3.4 Different layers of insulation and protective materials of an underground cable [42]. Where, SC—semiconductor, Cu—copper, Pb—lead, PE—poly ethylene

Table 3.3 Physical parameter data used for configuring the cables in the simulation model

Name of the cable	1000 R-Cu 130/225 (245) kV	800 R-Cu 130/225 (245) kV	400 R-Cu 130/225 (245) kV	500 R-Cu 76/132 (145) kV	400 R-Cu 76/132 (145) kV
Conductor diameter (mm)	38.2	34.7	23.3	26.7	23.2
Thickness of insulation (mm)	14.8	14.8	21.6	16.2	17.1
Al screen: out-side diameter of cable (mm)	85	82	85	77	75
Cu wire/Al sheath: out-side diameter of cable (mm)	88	84	87	79	77

Table 3.4 Resistivity of metals used in the cables

Material	Resistivity at 20 °C (mΩ)
Aluminum (Al)	2.803×10^{-8}
Copper (Cu)	1.724×10^{-8}
Lead (Pb)	21.4×10^{-8}

Table 3.5 Permittivity of insulation materials used in the cables

Material	Permittivity at 50 Hz
XLPE	2.35
Impregnated paper (high density)	3.8
Impregnated paper (low density)	3.3

Fig. 3.5 The cable that links Colombo Fort substation and Kelanitissa power station bus-bars, which appears in the transmission network of the simulation model of the power system of Sri Lanka

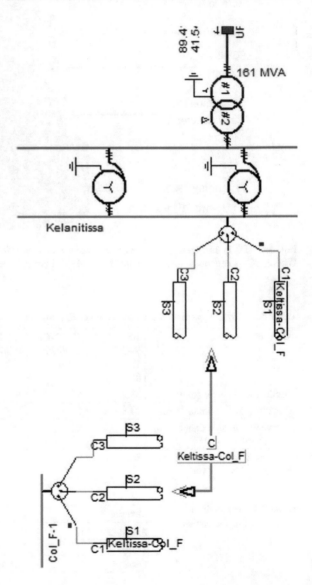

The resistivity of soil varies widely at different locations as the type of soil changes. At some locations, the soil can be very nonhomogeneous with multilayer soil structure, not far away from the earth surface. Often, there are several layers of

Fig. 3.6 PSCAD windows corresponding to power transmission cable from Kelanitissa to Colombo Fort substation named as 'Keltissa-Col_F' with it's **a** cable configuration window; **b** line model general data; **c** Bergeron model options

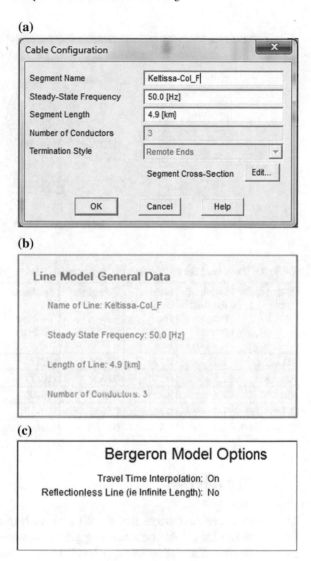

soil made up of loam, sand, clay, gravel, and rocks. The layers may be roughly horizontal to the surface or inclined at an angle to the surface. The resistivity also fluctuates seasonally due to changes in rainfall and temperature. The impact of temperature is only important near and below freezing point [44]. Hence it has been taken the (average) ground resistivity as 100 Ωm at a depth of 1.3 m from ground level, for the entire simulation.

Since there are several such cables cater for power transmission in the power system net-work, corresponding data which was used for cable configuration is available in Table 3.6.

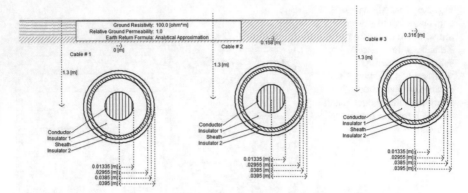

Fig. 3.7 132 kV under-ground cable lay-out corresponding to 'Keltissa-Col_F'

Table 3.6 Data used for configuring under-ground/submarine cables

Name of the cable	Depth of the cable from ground level (m)	Span between two cables (m)	Conductor diameter (m)	Thickness of insulator-1 (XLPE) (m)	Thickness of sheath (m)	Thickness of insulator-2 (m)
XLPE 150	1.3	0.174	0.0233	0.0216	9.25×10^{-3}	1×10^{-3}
Cu 350	1.3	0.154	0.0232	0.0171	8.8×10^{-3}	1×10^{-3}
Cu 500	1.3	0.158	0.0267	0.0162	8.825×10^{-3}	1×10^{-3}
XLPE 800	1.3	0.168	0.0347	0.0148	8.85×10^{-3}	1×10^{-3}
XLPE 1000	1.3	0.176	0.0382	0.0148	10^{-3}	1×10^{-3}

3.2.3 Transformers

Transformers are the devices that permit high voltage transmission and this drastically reduces losses. At corresponding grid substations, transmission voltage/s is step down to a distribution voltage level. Power transportation is accomplished through distribution systems. This includes small networks of radial or ring-main configuration and voltages are stepped down to end-user levels once again through transformers [14].

Transformer Types Available in the Power System
In this simulation generator transformers and transformers at grid sub-stations which step down 220–132 kV were considered. The generator transformers are with different capacities. The primary voltage is the voltage at which the electricity is generated. Some of the grid sub-stations are comprised with auto transformers that step down the voltage to 132 and 33 kV simultaneously. Since the intension of

the simulation is to model the transmission network, the 33 kV, which is at primary distribution level, was not considered.

Sample Calculation
The calculations were carried out referring to [45], pp. 36–37. Typical per-unit (pu) values of transformers are given in Tables 3.7 and 3.8. For the transformer capacity ranges given in Table 3.7, the leakage reactance (pu) on primary and secondary windings can be considered as equal for design purposes. In the same way, the copper loss on each side has been considered as equal.

Example:

Sample calculation of generator transformer parameters of Kotmale power station (less than 100 MVA):

Transformer rating = 90 MVA
Primary winding—star
Secondary winding—delta
Primary voltage—13.8 kV
Secondary voltage—220 kV

$$+ \text{ve sequence leakage reactance of primary}, x_1 = \left(\frac{+0.03}{100} \times 90 \right) + 0.03$$
$$= 0.057 \, \text{pu}$$

$$+ \text{ve sequence leakage reactance of secondary referring to primary}, x_1' = \left(\frac{13.8}{220} \right)^2 \times 0.057 \, \text{pu}$$
$$= 0.00393 \times 0.057 \, \text{pu}$$
$$= 0.0002 \, \text{pu}$$

$$\therefore + \text{ve sequence leakage reactance} = 0.057 + 0.0002 \, \text{pu}$$
$$= 0.057 \, \text{pu}$$

Table 3.7 Typical per-unit values of transformers [45]

Circuit element	Typical per-unit values	
	3–250 kVA	1–100 MVA
R_1 or R_2	0.009–0.005	0.005–0.002
X_{f1} or X_{f2}	0.008–0.025	0.03–0.06
X_m	20–30	50–200
R_m	20–50	100–500
I_0	0.05–0.03	0.02–0.005

Table 3.8 Per-unit values of transformer parameters [45]

S_n	kVA	1	10	100	1000	400,000
E_{np}	V	2400	2400	12470	69,000	13,800
E_{ns}	V	46	347	00	6900	424,000
I_{np}	A	0.417	4.17	8.02	14.5	29,000
I_{ns}	A	2.17	28.8	17	145	943
Z_{np}	Ω	5760	576	1555	4761	0.4761
Z_{ns}	Ω	211.6	12.0	3.60	47.61	449.4
R_1(pu)	–	0.0101	0.0090	0.0075	0.0057	0.00071
R_2(pu)	–	0.0090	0.0079	0.007	0.0053	0.00079
X_{f1}(pu)	–	0.0056	0.0075	0.0251	0.0317	0.0588
X_{f2}(pu)	–	0.0055	0.0075	0.0250	0.0315	0.06601
X_m(pu)	–	34.7	50.3	96.5	106	966
R_m(pu)	–	69.4	88.5	141.5	90.7	66
I_0(pu)	–	0.032	0.023	0.013	0.015	0.0018

$$\text{Copper loss resistance of primary, } r_1 = \left(\frac{-0.003}{100} \times 90 \right) + 0.005$$
$$= 0.0027 \, \text{pu}$$

$$\text{Copper loss resistance of secondary referring to primary, } r_1' = \left(\frac{13.8}{220} \right)^2 \times 0.0027 \, \text{pu}$$
$$= 0.00393 \times 0.0027 \, \text{pu}$$
$$= 0.000010 \, \text{pu}$$

$$\therefore \text{total copper loss resistance, } = 0.0027 + 0.00001 \, \text{pu}$$
$$= 0.00271 \, \text{pu}$$

$$\text{Air core reactance, } = 2 \times (+ve \, sequence \, reactance)$$
$$= 2 \times 0.057 \, \text{pu}$$
$$= 0.114 \, \text{pu}$$

The data given in Table 3.8 corresponds to transformer ratings from 1 kVA to 400 MVA. So calculation of transformer parameters whose ratings are above 100 MVA was carried out based on the data of the Table 3.8.

Sample calculation of parameters, of a transformer that has a rating above 100 MVA:

This is located at Biyagama-grid substation.
Transformer rating = 250 MVA
Transformer type—auto-transformer

Primary winding—star
Secondary winding—delta
Primary voltage—220 kV
Secondary voltage—132 kV

$$+\text{ve sequence leakage reactance of primary}, x_1 = \left(\frac{+0.0588 - 0.0317}{400}\right) \times 250 + 0.0317$$
$$= 0.04864 \, \text{pu}$$

$$+\text{ve sequence leakage reactance of secondary referring to primary}, x_2' = \left(\frac{220}{132}\right)^2 \times 0.04864 \, \text{pu}$$
$$= 0.1351 \, \text{pu}$$

$$\therefore \text{equivalent} + \text{ve sequence leakage reactance}, x_{eq} = x_1 + x_2' \, \text{pu}$$
$$= 0.0486 + 0.1351 \, \text{pu}$$
$$= 0.18375 \, \text{pu}$$

$$\text{Copper loss resistance of primary}, r_1 = \left(\frac{0.0007 - 0.0057}{400}\right) \times 250 + 0.0057$$
$$= 0.00258 \, \text{pu}$$

$$\text{Copper loss resistance of secondary referring to primary}, r_2' = \left(\frac{220}{132}\right)^2 \times 0.00258 \, \text{pu}$$
$$= 0.00717 \, \text{pu}$$

$$\therefore \text{total copper loss resistance}, = r_1 + r_2' \, \text{pu}$$
$$= 0.00258 + 0.00717 \, \text{pu}$$
$$= 0.00975 \, \text{pu}$$

$$\text{Air core reactance}, = 2 \times (+ \textit{ve sequence reactance})$$
$$= 2 \times 0.057 \, \text{pu}$$
$$= 0.114 \, \text{pu}$$

Values Used with PSCAD Window
There are three transformers each with a capacity of 90 MVA, installed in the Kotmale power station. Figure 3.8 shows them as they appear in the PSCAD simulation.

PSCAD windows corresponding to a transformer configuration and saturation for the generator transformers at Kotmale power plant are shown in Fig. 3.9a, b.

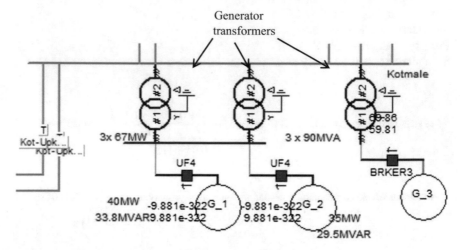

Fig. 3.8 Generator transformers located at the Kotmale power station, as they appear in the PSCAD simulation model

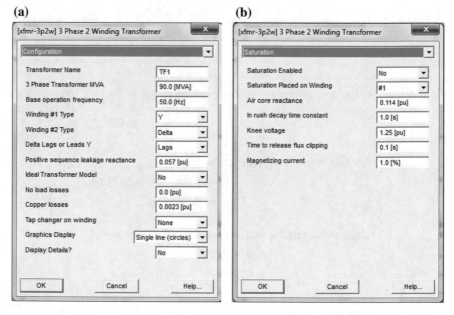

Fig. 3.9 PSCAD generator transformer windows of **a** configuration; **b** saturation; corresponding to the Kotmale power station

3.2.4 Generators

An electric generator is a device designed to take advantage of electromagnetic induction in order to convert movement into electricity. It is designed to obtain an induced current in a conductor (or set of conductors) as a result of mechanical movement, which is utilized to continually change a magnetic field near the conductor. The generator thus achieves a conversion of one physical form of energy into another—energy of motion into electrical energy—mediated by the magnetic field that exerts forces on the electric charges [46].

With reference to [6], synchronous generators can be roughly classified as shown in Fig. 3.10.

Further, a generating unit can be demonstrated as shown in the Fig. 3.11.

- Electricity is produced by a synchronous generator driven by a prime mover, usually a turbine or a diesel engine. The turbine is equipped with a turbine governor to control either the speed or the output power according to a preset power–frequency characteristic.
- Through a transmission network, the generated power is transmitted over to load-centres after stepping up the generated voltage (by a generator transformer).
- The DC excitation (or field) current, required to produce the magnetic field inside the generator, is provided by the exciter. The excitation current, and consequently the generator's terminal-voltage, is controlled by an automatic voltage regulator (AVR).

In modelling a real time generator, it is very important to have a clear idea about how the energy in mechanical form is converted to electrical form, in order to

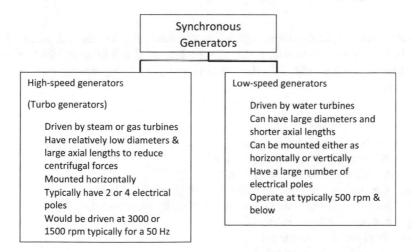

Fig. 3.10 Classification of synchronous generators referring to their speeds

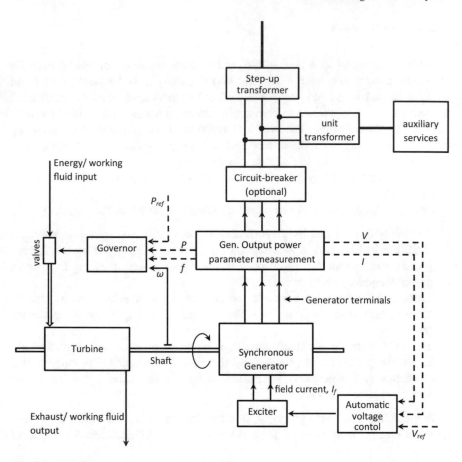

Fig. 3.11 Configuration of a generating unit [6]

maintain some specific frequency, while the electricity in demand varies. This would be very helpful in configuring the machine so that it behaves like a specific real-time generator. With reference to [11] and [13], basic generator model was developed.

Defining electrical-mechanical system terms:

ω rotational speed (rad/s)

α rotational acceleration

δ phase angle of a rotating machine

T_{net} net acceleration torque in a machine

T_{mech} mechanical torque exerted on the machine by the turbine

T_{elec} electrical torque exerted on the machine by the generator

P_{net} net accelerating power

P_{mec} mechanical power input

P_{elec} electrical power output

I moment of inertia for the machine

M angular momentum of the machine

- All these quantities (except phase angle) will be in per unit on the machine base, or in the case of ω, on the standard system frequency base.
- It is considered deviations of quantities about the steady state values.
- All steady-state or nominal values may be denoted with a sub-script '0' (e.g., ω_0, T_{net_0}).
- All deviations from nominal may be denoted by a 'Δ' (e.g., $\Delta\omega, \Delta T_{net}$).

$$I\alpha = T_{net}$$
$$M = \omega I \tag{3.1}$$
$$P_{net} = \omega T_{net} = \omega(I\alpha) = M\alpha$$

Consider a rotating machine with a steady speed of ω_0 and phase angle δ_0. Due to various electrical or mechanical disturbances, the machine will be subjected to differences in mechanical and electrical torque, causing it to accelerate or decelerate.

Consider a deviation of speed $\Delta\omega$ and a deviation in phase angle $\Delta\delta$, from nominal.

$$\omega = \omega_0 + \alpha t$$

$$\Delta\delta = \int (\omega_0 + \alpha t)dt - \int (\omega_0)dt$$

Machine absolute phase angle   Phase angle of reference axis

$$= \omega_0 t + \frac{1}{2}\alpha t^2 - \omega_0 t$$
$$= \frac{1}{2}\alpha t^2$$

Deviation from nominal speed,

$$\Delta\omega = \alpha t = \frac{d}{dt}(\Delta\delta)$$
$$\therefore T_{net} = I\alpha = I\frac{d}{dt}(\Delta\omega) = \frac{d^2}{dt^2}(\Delta\delta) \tag{3.2}$$

With reference to Appendix-C the relationship between net acceleration power and the electrical and mechanical power is:

$$P_{net} = P_{mech} - P_{elec}$$

This can be written as a sum of the steady-state value and the deviation term,

$$P_{net} = P_{net_0} - \Delta P_{net}$$

where

$$P_{net_0} = P_{mech_0} - P_{elec_0}$$
$$\Delta P_{net} = \Delta P_{mech} - \Delta P_{elec}$$

Then,

$$P_{net} = (P_{mech_0} - P_{elec_0}) + (\Delta P_{mech} - \Delta P_{elec}) \tag{3.3}$$

Similarly for torques,

$$T_{net} = (T_{mech_0} - T_{elec_0}) + (\Delta T_{mech} - \Delta T_{elec}) \tag{3.4}$$

Since Eq. (3.1) $\Rightarrow$,

$$P_{net} = \omega T_{net}$$

$$P_{net} = P_{net_0} + \Delta P_{net} = (\omega_0 + \Delta\omega)(T_{net_0} + \Delta T_{net})$$

From Eqs. (3.3) and (3.4)

$$(P_{mech_0} - P_{elec_0}) + (\Delta P_{mech} - \Delta P_{elec}) = (\omega_0 + \Delta\omega)[(T_{mech_0} - T_{elec_0}) + (\Delta T_{mech} - \Delta T_{elec})]$$

Assume that the steady-state quantities can be factored out since

$$P_{mech_0} = P_{elec_0}$$

and

$$T_{mech_0} = T_{elec_0}$$

and further assume that the second-order terms involving products of $\Delta\omega$ with ΔT_{mech} and ΔT_{elec} can be neglected. Then,

$$(\Delta P_{mech} - \Delta P_{elec}) = \omega_0(\Delta T_{mech} - \Delta T_{elec}) \tag{3.5}$$

As shown in Eq. (3.2), the net torque is related to the speed change as follows:

$$(T_{mech_0} - T_{elec_0}) + (\Delta T_{mech} - \Delta T_{elec}) = I\frac{d}{dt}(\Delta\omega) \qquad (3.6)$$

Then, since $T_{mech0} = T_{elec0}$, by combining Eqs. (3.5) and (3.6),

$$(\Delta P_{mech} - \Delta P_{elec}) = \omega_0 I\frac{d}{dt}(\Delta\omega)$$

$$= M\frac{d}{dt}(\Delta\omega)$$

$$\mathcal{L}(\Delta P_{mech} - \Delta P_{elec}) = \mathcal{L}\left(M\frac{d}{dt}(\Delta\omega)\right)$$

∴ In Laplace transform operator notation,

$$\Delta P_{mech} - \Delta P_{elec} = Ms(\Delta\omega)$$

This relationship between mechanical and electrical power and speed change can be given in a block-diagram as shown in Fig. 3.12.

The loads on a power system consist of a variety of electrical devices such as,

- purely resistive
- motor loads with variable power–frequency characteristics

Since motor loads are a dominant part of the electrical load, the effect of the change in load due to the change in frequency can be given by:

$$\Delta P_{L(freq)} = D\Delta\omega \, or \, D = \frac{\Delta P_{L(freq)}}{\Delta\omega}$$

∴ The net change in P_{elec} in Fig. 3.12, (explains in Eq. (3.6)) is:

$$\Delta P_{elec} = \underbrace{\Delta P_L}_{Nonfreqency-sensitive\,load\,change} + \underbrace{D\Delta\omega}_{Freqency-sensitive\,load\,change}$$

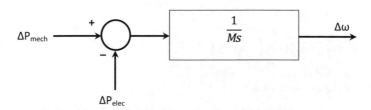

Fig. 3.12 Relationship between mechanical and electrical power and speed change

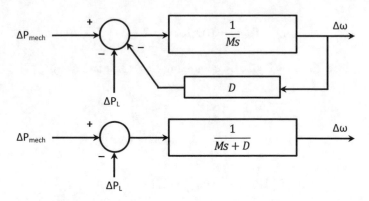

Fig. 3.13 Block diagrams demonstrating the effect of change in frequency sensitive and non-frequency sensitive load

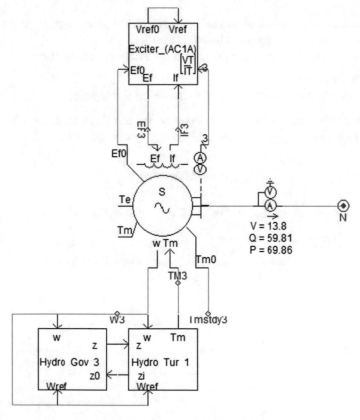

Fig. 3.14 A generator unit at Kotmale power station with its hydro turbine, governor and exciter

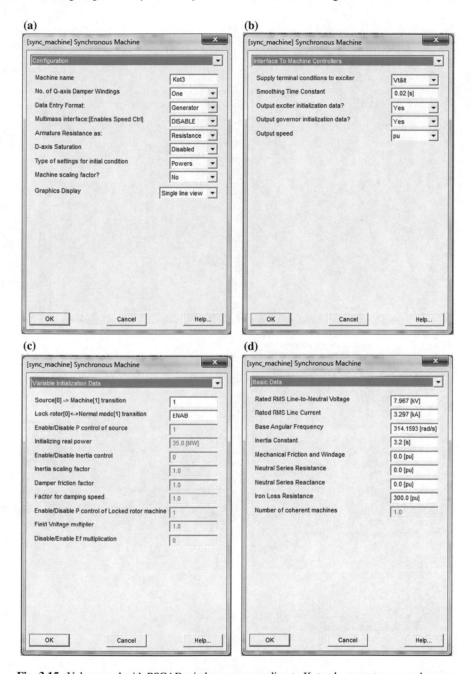

Fig. 3.15 Values used with PSCAD window corresponding to Kotmale generator; **a** synchronous machine configuration; **b** interface to machine controllers; **c** variable initialization data; **d** basic data; **e** initial conditions; **f** initial conditions if starting as a machine; **g** output variable names and **h** power output of the simulated generator 'Kotmale-3'

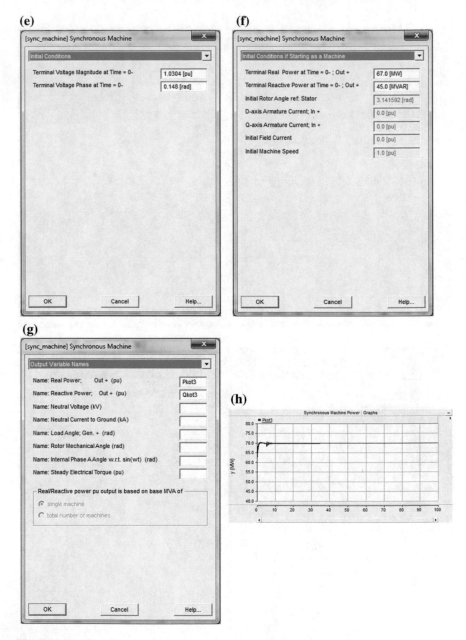

Fig. 3.15 (continued)

Table 3.9 Inertia constants of different types of generators

Generator type	Range of inertia constant (s)
Hydraulic unit (water wheel generator)	2–4
Thermal unit:	
• Condensing (1800 rpm)	9–6
• Condensing (3000 rpm)	7–4
• Non-condensing (3000 rpm)	4–3

This describes the block-diagram shown in the Fig. 3.13.

Generator Types Used in Modeling the Power System
In this power system simulation model, basically two types of generators were used. They are:

- Hydro-generators
- Steam-turbine generators

All the thermal power generators which were committed at the instant of the corresponding load-flow simulation were represented with steam turbine generators.

Simplified Schematic Diagram and Corresponding Control System
With reference to the PSCAD simulation model, the Fig. 3.14 shows a hydro-generator unit located at Kotmale power station, (Kotmale3) with its corresponding governor, turbine and exciter units.

Values Used with PSCAD Window
Figure 3.15 with (a) Synchronous machine configuration, (b) Interface to machine controllers, (c) Variable initialization data, (d) Basic data, (e) Initial conditions, (f) Initial conditions if starting as a machine, (g) Output variable names and (h) power output of the simulated generator 'Kotmale-3' show the configuration windows and graph of power generated, corresponding to Kotmale hydro-generator.

Sample Calculation
Sample calculation of parameters, of the generator Kotmale3:

Rated (active) power = 67 MW
Rated (reactive) power = 78.8 MVA
Rated output voltage = 13.8 kV

$$\text{Rated RMS line to neutral voltage} = \frac{13.8}{\sqrt{3}} = 7.967\,\text{kV}$$

$$\text{Rated RMS line current} = \frac{78.8 \times 10^6}{\sqrt{3} \times 13.8 \times 10^3} = 3.297\,\text{kA}$$

Table 3.10 Inertia constants
of generators obtained from
the CEB

Station/generator	Inertia constant (H s)
Barge	1.62
KCCPgas	8
KCCPsteam	4
Sapugaskanda-1	3.2
Sapugaskanda-2	3.2
Sapugaskanda-3	3.2
Kerawalapitiya-1	–
Kerawalapitiya-2	–
Kerawalapitiya-3	–
KHD (Asia power)-1	0.997
KHD (Asia power)-2	0.997
Puttalam Coal	–
Upper Kotmale	–
Heledanavi-1	1.3
Heledanavi-2	1.3
Kotmale-3	3.02
Polpitiya-1	2.84
Polpitiya-2	2.84
Canyon-1	3.8
Canyon-2	3.8
Kukule-1	3
Kukule-2	3
New Laxapana-1	3.3
New Laxapana-2	3.3
Victoria-1	3.45
Victoria-2	3.45
Victoria-3	3.45
Ukuwela	2.72
Randenigala	3.65
Wimalasurendra	3.2
Laxapana-1	3.17
Laxapana-2	2.45
Samanalawewa-1	3.4
Samanalawewa-2	3.4
Bowatenna	4
Rantembe-1	2.62
Embilipitiya	1.1

Table 3.11 Inertia constants used for the units considered in the simulation

Station/generator	Unit capacity (MW)	Generator type	Inertia constant (H s)
Barge	60	Thermal (diesel)	3.14
KCCPgas	161	Thermal (gas)	3.79
KCCPsteam	81	Thermal (steam)	3.14
Sapugaskanda-1	9 MW × 4	Thermal (diesel: coherent—4 units.)	2.53
Sapugaskanda-2	9 MW × 4	Thermal (diesel: coherent—4 units.)	2.53
Sapugaskanda-3	18 MW × 4	Thermal (diesel: coherent—4 units.)	2.65
Kerawalapitiya-1	100 MW	Thermal (gas)	3.62
Kerawalapitiya-2	100 MW	Thermal (gas)	3.62
Kerawalapitiya-3	100 MW	Thermal (gas)	3.62
KHD (Asia power)-1	6.375 MW × 4	Thermal (diesel: coherent—4 units.)	2.51
KHD (Asia power)-2	6.375 MW × 4	Thermal (diesel: coherent—4 units.)	2.51
Puttalam Coal	300 MW	Thermal (steam)	4
Upper kotmale	75 MW	Hydro	3.47
Heledanavi-1	17 MW × 3	Thermal (diesel: coherent—3 units.)	2.63
Heledanavi-2	17 MW × 3	Thermal (diesel: coherent—3 units.)	2.63
Kotmale-3	67 MW	Hydro	3.2
Polpitiya-1	37.5 MW	Hydro	2.84
Polpitiya-2	37.5 MW	Hydro	2.84
Canyon-1	30 MW	Hydro	3.8
Canyon-2	30 MW	Hydro	3.8
Kukule-1	35 MW	Hydro	3.0
Kukule-2	35 MW	Hydro	3.0
New Laxapana-1	50 MW	Hydro	3.3
New Laxapana-2	50 MW	Hydro	3.3
Victoria-1	70 MW	Hydro	3.45
Victoria-2	70 MW	Hydro	3.45
Victoria-3	70 MW	Hydro	3.45
Ukuwela	20 MW	Hydro	2.72
Randenigala	61 MW	Hydro	3.65
Wimalasurendra	25 MW	Hydro	3.2
Laxapana-1	8.33 MW	Hydro (coherent—3 units.)	3.17
Laxapana-2	12.5 MW	Hydro (coherent—2 units.)	2.45
Samanalawewa-1	60 MW	Hydro	3.4

(continued)

Table 3.11 (continued)

Station/generator	Unit capacity (MW)	Generator type	Inertia constant (H s)
Samanalawewa-2	60 MW	Hydro	3.4
Bowatenna	40 MW	Hydro	3.0
Rantembe-1	25 MW	Hydro	2.62
Embilipitiya	100 MW	Thermal (diesel)	3.62

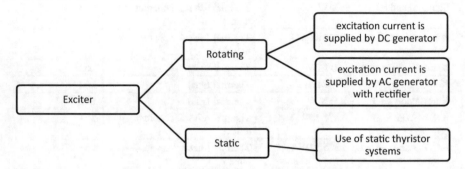

Fig. 3.16 Basic types of exciters

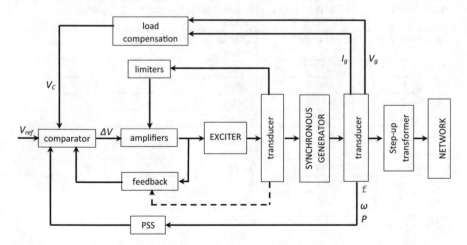

Fig. 3.17 Block diagram of the excitation and AVR system. PSS, power system stabilizer, [6]

$$\text{Base angular frequency} = 2\pi f = 2 \times \pi \times 50 \, \text{rad/s} = 314.16 \, \text{rad/s}$$

Finding inertia constant:

With reference to [11], the inertia constants H of synchronous generators (hydro generator and steam turbine generator) can be briefed as shown in Table 3.9. Based on these data as well as the data obtained from the CEB (shown in Table 3.10), the inertia constants used for the simulation were calculated.

Where range is given, the 1st figure applies to the smaller MVA.

$$\text{Inertia constant (calculated) of Kotmale3 generator} = \left(\frac{3.17 - 3.45}{8.3 - 71}\right) \times (P - 8.3) + 3.17$$

$$= \left(\frac{3.17 - 3.45}{8.3 - 71}\right) \times (67 - 8.3) + 3.17$$

$$= 3.432\,\text{s}$$

$$\text{But the experimental value} = 3.432\,\text{s}$$

$$\text{Hence, the average} = (3.432 + 3.02)/2$$
$$= 6.45/2\,\text{s}$$
$$\sim 3.2\,\text{s}$$

The inertia constants used for the corresponding units that were committed at the instant of the load-flow concerned are given in Table 3.11. Even though it has been assumed that all thermal power generators in the power system are steam-turbine power plants, in calculating inertia constants corresponding to gas-turbine/diesel power plants, they were calculated considering them as non-condensing turbine generators with inertias in the range of 4–3 s. Because of the doubts with the accuracy of experimental values obtained, in certain cases approximated values were used.

3.2.5   Exciters

The generator excitation system consists of an exciter and an AVR and is necessary to supply the generator with DC field current as shown in Fig. 3.16. The AVR regulates the generator terminal voltage by controlling the amount of current

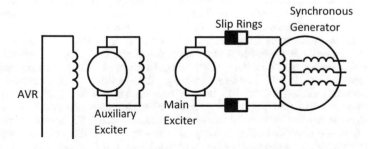

Fig. 3.18   Cascaded DC generators. AVR, automatic voltage regulator [6]

Fig. 3.19 Exciter model 'AC1A' which was used for the simulation

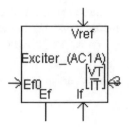

supplied to the generator field winding by the exciter. Generally the exciters can be classified as shown in Fig. 3.16.

Figure 3.17 demonstrates how the AVR subsystem operates [6]. The generator terminal voltage, V_g is measured and compensated for the load current I_g and compared with the desired reference voltage to get the voltage error ΔV. This error is the amplified and used to alter the exciter output, and consequently the generator field current, so that the voltage error is eliminated. This represents a closed loop

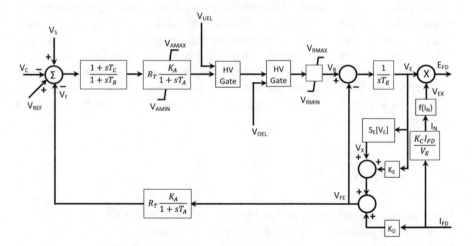

Fig. 3.20 IEEE Alternator Supplied Rectifier Excitation System #1 (AC1A) as in PSCAD. Where, Vc—output of terminal voltage and load compensation elements (pu); V_{REF}—voltage regulator reference (determined to satisfy initial conditions) (pu); V_S—combined power system stabilizer and possibly discontinuous control output after any limits or switching, as assumed with terminal voltage and reference signals (pu); V_P—excitation system stabilizer output [pu]; V_{AMAX}, V_{AMIN}—maximum and minimum regulator output limits [pu]; V_R—voltage regulator output (pu); V_{FE}—signal proportional to exciter field current; T_E—exciter time constant, integration rate associated with exciter control (s); V_E—exciter voltage back of commutating reactance (pu); K_E—exciter constant related to self-excited field (pu); K_D—demagnetizing factor, a function value at the corresponding exciter voltage, VE, back of commutation reactance (pu); K_F—excitation control system stabilizer gain [pu]; V_X—signal proportional to exciter saturation (pu); S_E [V_E]—exciter saturation function value at the corresponding exciter voltage, VE, back of commutation reactance (pu); K_C—rectifier loading factor proportional to commutation reactance (pu); I_{FD}—synchronous machine field current (pu); I_N—normalized exciter load current (pu); F_{EX}—rectifier loading factor, a function of I_N (pu); E_{FD}—exciter output voltage (pu)

control system. The regulation process is stabilized using a negative feedback loop taken directly from either the amplifier or the exciter.

Exciter Types Used in Modeling the Power System
DC generators usually have relatively low power ratings; they are cascaded to obtain the necessary output as shown in Fig. 3.18.

Because of commutation problems with DC generators, this type of exciter cannot be used for large generators which require large excitation currents (usually power rating of the exciter is in the range 0.2–0.8% of the generator's megawatt rating) [6]. Hence, in this power system simulation, for all generators, exciter model AC1A (Figs. 3.14 and 3.19), which is an ac-exciter was used.

Simplified Schematic Diagram and Corresponding Control System
Figure 3.19 shows the exciter model 'AC1A' which was used for the simulation. The control system of the 'IEEE Alternator Supplied Rectifier Excitation System #1 (AC1A)' as given in PSCAD is shown in Fig. 3.20. Further, with reference to [6], this can be considered as 'the excitation system with AC alternator and the uncontrolled rectifier' given in Fig. 3.21. Referring to Chap. 11 of [6], the corresponding values for the parameters of the AC exciter were decided. There it has been justifies that the same DC-exciter parameters could be used for the above AC-exciter system. Therefore the corresponding parameter values were decided accordingly.

The excitation system is stabilized by the feedback loop with transfer function,

$$K_G(s) = \frac{K_F s}{(1 + T_F s)}$$

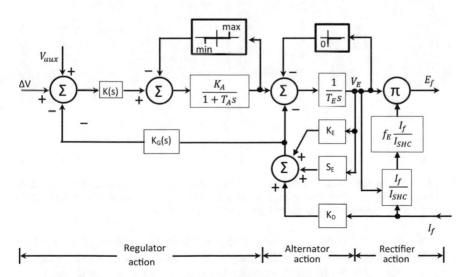

Fig. 3.21 The excitation system with AC alternator and the uncontrolled rectifier [6]

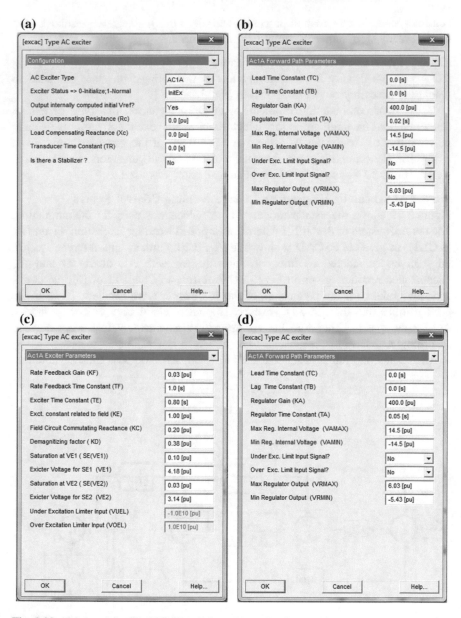

Fig. 3.22 Values used with PSCAD windows in configuring Ac1A exciters; **a** configuration, **b** hydro-gen.'s forward path parameters, **c** hydro-gen.'s exciter parameters, **d** steam turbine gen.'s forward path parameters, **e** steam turbine gen.'s exciter parameters

Alternatively the excitation system could be stabilized by supplying this block directly from the output of the voltage regulator or from the excitation voltage E_f.

In the Fig. 3.21, the feedback stabilization is done by an additional block with the transfer function K(s) in the forward path preceding the regulator block. Both

(e)

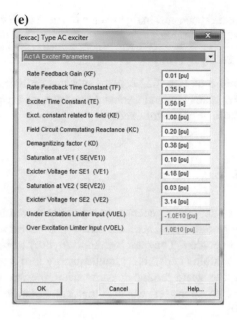

[excac] Type AC exciter   x

Ac1A Exciter Parameters   ▼

Rate Feedback Gain (KF)	0.01 [pu]
Rate Feedback Time Constant (TF)	0.35 [s]
Exciter Time Constant (TE)	0.50 [s]
Exct. constant related to field (KE)	1.00 [pu]
Field Circuit Commutating Reactance (KC)	0.20 [pu]
Demagnitizing factor (KD)	0.38 [pu]
Saturation at VE1 (SE(VE1))	0.10 [pu]
Exicter Voltage for SE1 (VE1)	4.18 [pu]
Saturation at VE2 (SE(VE2))	0.03 [pu]
Exciter Voltage for SE2 (VE2)	3.14 [pu]
Under Excitation Limiter Input (VUEL)	-1.0E10 [pu]
Over Excitation Limiter Input (VOEL)	1.0E10 [pu]

| OK | Cancel | Help... |

Fig. 3.22 (continued)

$K_G(s)$ and $K(s)$ depend on the specific excitation system and can be implemented by either analogue or digital techniques. $K(s)$ can be given as the transfer function,

$$K(s) = \frac{(1 + T_C s)}{(1 + T_B s)}$$

This can be simplified by neglecting T_C and T_B variables and thereby $K(s) = 1$.

With reference to [6] page 465, A (separately excited) DC exciter usually operates so that the parameter $K_E = 0.8$–0.95. This can be approximated to $K_E = 1$. The time constant, $T_E < 1$ s,

$\therefore T_E \approx 0.5$ s. Typical values for time constant $T_A = 0.05$–0.2 s and gain $K_A = 20$–400. The high regulator gain is necessary to ensure small voltage regulation of the order of 0.5%. Although this high gain ensures low steady-state error, when coupled with the length of the time constants the transient performance of the exciter is unsatisfactory. To achieve an acceptable transient performance the system must be stabilized in some way that reduces the transient (high-frequency) gain. This is achieved by a feedback stabilization signal represented by the first-order differentiating element with gain K_F and time constant T_F. Typical values of the parameters in this element are $T_F = 0.35$–1 s and $K_F = 0.01$–0.1.

Sample Calculation

Referring to Sect. 3.2.6.2—Simplified schematic diagram and corresponding control system, Ac1A forward path parameters and exciter parameters were set. As given in Fig. 3.22, most of the parameters in control systems of 'hydro-power generator exciter' and 'steam turbine generator exciter' are common to both, but few are different from each other.

Values Used with PSCAD Window

Figures *a, b, c, d* and *e* of Fig. 3.22 corresponds to Configuration and respective Ac1A Forward Path Parameters and Exciter Parameters of Hydro-power generator exciter and Steam turbine generator exciter.

3.2.6 Turbines

In a power system, synchronous generators are usually driven by steam turbines or gas turbines or hydro-turbines. There are two types of prime movers used for large-scale power generation. They are hydraulic (hydro) turbines, and steam turbines. The hydraulic turbine converts hydraulic energy into rotating kinetic energy of the prime mover, the steam turbine converts steam thermal energy into rotating kinetic energy of the prime mover; which is then converted into electric power by the generator [6, 40].

Turbine Types Used in Modeling the Power System
As stated in previous chapters all thermal power plants in the power system were assumed as steam turbine power plants. Therefore,

- steam turbine and
- hydro turbine

are the two types of turbines considered in this power system modelling.

Steam Turbines
In coal-burn or oil-burn power plants, the energy contained in the fuel is used to produce high-pressure, high-temperature steam in the boiler. The energy in the steam is then converted to mechanical energy in axial flow steam turbines. Each turbine consists of a number of stationary and rotating blades concentrated into groups/stages. Typically a complete steam turbine can be divided into three or more stages, with each turbine stage being connected in *tandem* on a common shaft. Dividing the turbine into stages in this way allows the steam to be reheated between stages to increase its enthalpy and consequently increase the overall efficiency of the steam cycle. Modern coal-fired steam turbines have thermal efficiency reaching 45% [6, 11]. Steam turbines can be classified according to the number of re-heat units it has.

- Non-reheat systems
- Single reheat systems
- Double reheat systems

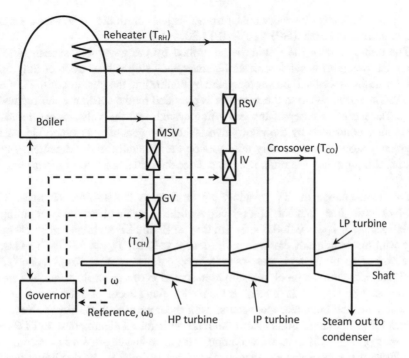

Fig. 3.23 Configuration of a tandem compound single-reheat steam turbine, [6]

Simplified Schematic Diagram and Corresponding Control System of Steam Turbine

A steam turbine configuration with a single tandem reheat arrangement is shown in Fig. 3.23. Generally a turbine has three stages in three sections,

- High pressure (HP)
- Intermediate pressure (IP)
- Low pressure (LP)

Steam from the boiler enters the steam chest and flows through the main emergency stop valve (MSV) and governor control valve (GV) to the HP turbine. In this HP turbine the steam is partially expanded (i.e. pressure is slightly reduced). Then it is directed back to the boiler to be reheated in the heat-exchanger to increase its enthalpy. The steam then flows through the reheat emergency stop valve (RSV) and the intercept control valve (IV) to the IP turbine where it is again expanded and made to do work. This IV is in use only for rapid control of turbine mechanical power during an over-speed,

- Ahead of the re-heater
- Controls steam flow to IP and LP sections.

On leaving the IP stage the steam flows through the crossover piping for final expansion in the LP turbine. Finally the steam flows to the condenser to complete the cycle. Typically the individual turbine stages contribute to the total turbine torque in the ratio 30% (HP) : 40% (IP) : 30% (LP).

The steam flow in the turbine is controlled by the governing system (GOV). When the generator is synchronized the emergency stop valves are kept fully open and the turbine speed and power regulated by controlling the position of the GV and the IV. The speed signal to the governor is provided by the speed measuring device (SD). The main amplifier of the governing system and the valve mover is an oil servomotor controlled by the pilot valve. When the generator is synchronized the emergency stop valves are only used to stop the generator under emergency conditions, although they are often used to control the initial start-up of the turbine [6, 11].

Due to a change in GV opening, steam flow to the turbines changes. This involves some time constant, T_{CH} (∵charging time of steam chest and inlet piping to the HP section). $T_{CH} = 0.2–0.3$ s. Steam flow in IP and LP sections can be changed only with the buildup of pressure in the reheat volume. $T_{RH} = 5–10$ s. The steam flow from IP to LP through cross-over involves a time constant, T_{CO}. $T_{CO} \approx 0.5$ s.

Figure 3.24 shows a block diagram representation of a control system of a steam turbine as per [11]. This model accounts for the effects of inlet steam chest, re-heater, and the nonlinear characteristics of the control and intercept valves. The re-heater representation differs from the representation of steam chest and LP inlet crossover piping. This is to allow computation of re-heater pressure to account for the effects of intercept valve actuation. 'Base power equal to the maximum turbine power at rated main steam pressure with the control valves fully open' can be considered as a convenient per unit system. In this system, CV position is 1.0 pu when fully open. Then,

$$F_{HP} + F_{IP} + F_{LP} = 1.0\,\text{pu}$$

Figure 3.24 shows a single reheat tandem-compound steam turbine model with its (a) turbine configuration and (b) block diagram representation [11].

Comparing the control systems given in PSCAD (Fig. 3.25) and [11] (Fig. 3.26) for a steam turbine, the corresponding time constants and gains were identified. The typical values given in [11] were used in configuring the steam turbine in the power system simulation model (Table 3.14).

It has been assumed that no cross over piping is available for the turbine system. Hence there is no T_{CO}. ∵ $T_{CO1} = T_{CO2} = 0.0$ s. A set of typical values as given in [11] are appeared in Table 3.12. Their ranges of existence were discussed in detail in this chapter previously.

Referring to the typical values given in [11] and Table 3.12, values for the parameters in the control system of 'Generic turbine Model including intercept valve effect as given in PSCAD' were decided. Corresponding values are appeared in Table 3.13. The same turbine model configuration was used for all thermal power plants in the PSCAD simulation model (of the power system of Sri Lanka) (Table 3.14).

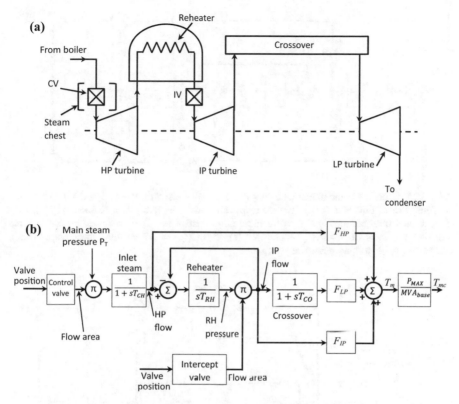

Fig. 3.24 Control system of a single reheat tandem compound steam turbine as per [11]; **a** turbine configuration; **b** block diagram representation. Where, T_{CH} = time constant of main inlet volumes and steam chest; T_{RH} = time constant of re-heater; T_{CO} = time constant of crossover piping and LP inlet volumes; T_m = total turbine torque in per unit of maximum turbine power; T_{mc} = total turbine mechanical torque in per unit of common MVA base; P_{MAX} = maximum turbine power in MW; F_{HP}, F_{IP}, F_{LP} = fraction of total turbine power generated by HP, IP, LP sections, respectively; MVA_{base} = common MVA base

Values Used with PSCAD Window

Figures *a, b, c, d* and *e* of Fig. 3.27 corresponds to actual parameter values used in 'Generic Turbine Model including Intercept Valve effect (TUR2)' Configuration, Hp Turbine: Contributions, Lp Turbine: Contributions, Time Constants and Intercept Valve.

This Steam_Tur_2 models an IEEE type thermal turbine and Fig. 3.28 shows that thermal (steam) turbine model—Generic turbine model including intercept valve effect (TUR2), which was used for the simulation.

Hydro Turbines

For hydro power plants the hydro turbine model 'Non-elastic water column without surge tank (TUR1)' was used. Figure 3.29 shows the above model as it appears in the PSCAD simulation model.

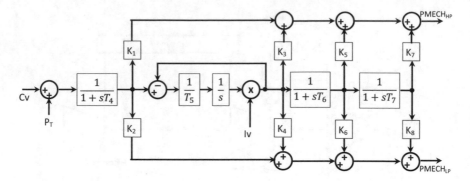

Fig. 3.25 Generic turbine model including intercept valve effect as given in PSCAD. Where, Cv = Control valve flow area (p.u.), Iv = Intercept valve flow area (p.u.), K = K fraction (p.u.), T_4 = Steam chest time constant (s), T_5 = Reheater time constant (s), T_6 = Reheater/ cross-over time constant (s), T_7 = Cross-over time constant (s), $PMECH_{HP}$ = HP turbine power output (p.u.), $PMECH_{LP}$ = LP turbine power output (p.u.)

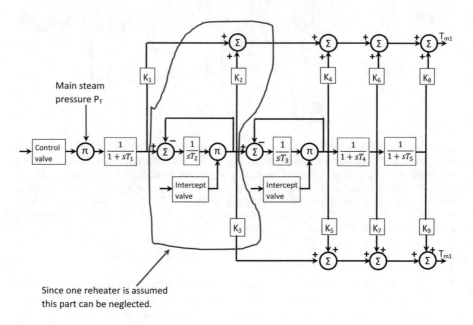

Fig. 3.26 Generic model for steam turbines [11]

The corresponding block diagram of the control system of the 'Non-elastic water column without surge tank (TUR1)' is appeared in Fig. 3.30.

Values Used with PSCAD Window

In Fig. 3.31 the parameter values with PSCAD configuration is given.

Table 3.12 Typical values in Fig. 3.26 corresponding to Fig. 3.25

Gain, K or Time-constant, T in Fig. 3.26	Corresponding values in Fig. 3.25
T_1	T_{CH}
T_2	neglected
T_3	T_{RH}
T_4	T_{CO1} ($-$ neglected)
T_5	T_{CO2} ($-$ neglected)
K_1	$=F_{HP}$
K_2	$=0$
K_3	$=0$
K_4	$=0$
K_5	F_{IP}
K_6	F_{LP1}
K_7	$=0$
K_8	$=0$
K_9	F_{LP2}

Table 3.13 Comparison of time constants and gains in Fig. 3.24 (as per [3]-Kundur's) and Fig. 3.25 (PSCAD)

Kundur's	PSCAD
T_1	$T_4 = T_{CH}$
T_3	$T_5 = T_{RH}$
T_4	$T_6 = T_{CO1}$
T_5	$T_7 = T_{CO2}$
K_1	$K_1 = F_{HP}$
K_4	K_3
K_5	$K_4 = F_{IP}$
K_6	$K_5 = F_{LP1}$
K_7	K_6
K_8	K_7
K_9	$K_8 = F_{LP2}$

Table 3.14 Values used for the parameters in the control system of 'Generic turbine Model including intercept valve effect as given in PSCAD'

PSCAD	Values as per the Fig. 3.27
T_4	$T_{CH} = 0.3$ s
T_5	$T_{RH} = 5.0$ s
T_6	$T_{CO1} = 0.0$ s
T_7	$T_{CO2} = 0.0$ s
K_1	$F_{HP} = 0.297$
K_2	$=0.001$
K_3	$=0.001$
K_4	$F_{IP} = 0.697$
K_5	$F_{LP1} = 0.001$
K_6	$=0.001$
K_7	$=0.001$
K_8	$F_{LP2} = 0$

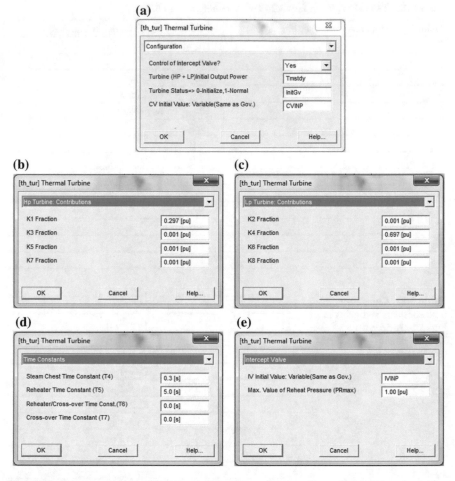

(a)

[th_tur] Thermal Turbine

Configuration

Control of Intercept Valve?   Yes

Turbine (HP + LP)Initial Output Power   Tmstdy

Turbine Status=> 0-Initialize,1-Normal   InitGv

CV Initial Value: Variable(Same as Gov.)   CVINP

OK   Cancel   Help...

(b)

[th_tur] Thermal Turbine

Hp Turbine: Contributions

K1 Fraction   0.297 [pu]

K3 Fraction   0.001 [pu]

K5 Fraction   0.001 [pu]

K7 Fraction   0.001 [pu]

OK   Cancel   Help...

(c)

[th_tur] Thermal Turbine

Lp Turbine: Contributions

K2 Fraction   0.001 [pu]

K4 Fraction   0.697 [pu]

K6 Fraction   0.001 [pu]

K8 Fraction   0.001 [pu]

OK   Cancel   Help...

(d)

[th_tur] Thermal Turbine

Time Constants

Steam Chest Time Constant (T4)   0.3 [s]

Reheater Time Constant (T5)   5.0 [s]

Reheater/Cross-over Time Const.(T6)   0.0 [s]

Cross-over Time Constant (T7)   0.0 [s]

OK   Cancel   Help...

(e)

[th_tur] Thermal Turbine

Intercept Valve

IV Initial Value: Variable(Same as Gov.)   IVINP

Max. Value of Reheat Pressure (PRmax)   1.00 [pu]

OK   Cancel   Help...

Fig. 3.27 Values used with PSCAD windows in configuring Steam_Tur_2 (Generic turbine model including intercept valve effect); **a** configuration, **b** Hp turbine: contributions, **c** Lp turbine: contributions, **d** time constants, **e** intercept valve

Fig. 3.28 'Steam_Tur_2' Generic turbine model including intercept valve effect (TUR2) used for the power system simulation

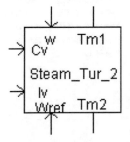

Fig. 3.29 Hydro turbine model used in power system simulation—'Non-elastic water column without surge tank (TUR1)'

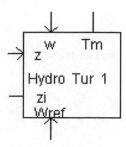

3.2.7   Governors

Each turbine is equipped with a governing system (Fig. 3.11) to provide a mechanism by which the turbine can be started, run up to the operating speed and operate on load with the required power output [6, 40].

Governor Types Used in Modeling the Power System
For the steam turbines and hydro turbines in the simulation,

- steam governors and
- hydro governors

were used respectively. GOV2: Mechanical-Hydraulic Controls (GE) and GOV3: Electro-Hydraulic Controls (GE) were used in simulating steam turbine governors. For the control of hydro turbines, governor models, GOV1: Mechanical-Hydraulic Controls and GOV3: Enhanced Controls for Load Rejection Studies were used. Table 3.15 gives the Governor control systems of the PSCAD simulation with corresponding turbines.

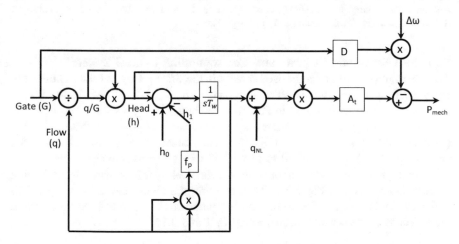

Fig. 3.30 Block diagram of control system of 'Non-elastic water column without surge tank', TUR1, PSCAD simulation model

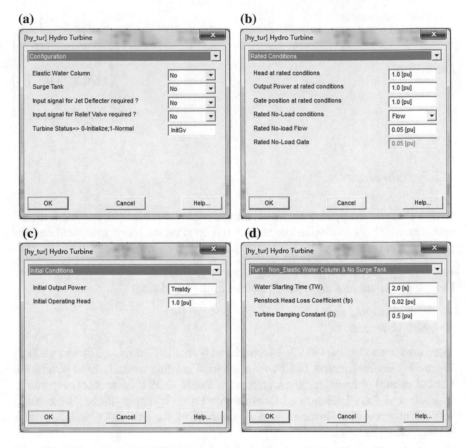

Fig. 3.31 Values used with PSCAD windows in configuring Hydro_Tur_1 (Non-elastic water column without surge tank (TUR1)); **a** configuration, **b** rated conditions, **c** initial conditions, **d** Tur: Non_Elastic Water Column and No Surge Tank

Simplified Schematic Diagram and Corresponding Control System

The Thermal (steam) governor used mostly is the Steam_Gov_2 (GOV2) but only for modelling Sapugaskanda-PP the Steam_Gov_3 (GOV3) was used. Hydro Governor-2 and 3 as appeared in PSCAD are shown in Fig. 3.32a, b respectively.

The block diagrams of the control systems of thermal governor 'Steam_Gov_2 (Mechanical-hydraulic controls (GOV2))' and 'Steam_Gov_3 (Electro-hydraulic controls (GOV3))' are appeared in Figs. 3.33 and 3.35 respectively.

As per [11], the block diagram of 'Mechanical-hydraulic controls governing system' is appeared in Fig. 3.32. By comparing the block diagrams in Figs. 3.33 and 3.34, typical values for corresponding parameters of PSCAD control system were decided. These values are appeared in Table 3.16.

Table 3.15 Governor Models of PSCAD used in controlling steam and hydro turbines of the simulation of the Power system of Sri Lanka

Thermal (steam) Governors		Hydro Governors	
Barge	GOV2	Upper Kotmale	GOV1[a]
KHD-1	GOV2	Kotmale-1	GOV3
KHD-2	GOV2	Kotmale-2	GOV3
Keravalapitiya-1	GOV2	Kotmale-3	GOV3
Keravalapitiya-2	GOV2	Kukule-1	GOV3
Keravalapitiya-3	GOV2	Kukule-2	GOV3
Kelanitissa-1 (CombinedCyclePP)	GOV2	Samanalawewa-1	GOV3
Kelanitissa-2 (CombinedCyclePP)	GOV2	Samanalawewa-2	GOV3
Sapugaskanda-1	GOV3[b]	Canyon-1	GOV3
Sapugaskanda-2	GOV3[b]	Canyon-2	GOV3
Sapugaskanda-3	GOV3[b]	New Laxapana-1	GOV3
Coal Power Plant	GOV2	New Laxapana-2	GOV3
Heladanavi-1	GOV2	Laxapana-1	GOV3
Heladanavi-2	GOV2	Laxapana-2	GOV3
Embilipitiya	GOV2	Polpitiya-1	GOV3
		Polpitiya-2	GOV3
		Victoria-1	GOV3
		Victoria-2	GOV3
		Victoria-3	GOV3
		Wimalasurendra	GOV3
		Randenigala	GOV3
		Ukuwela	GOV3
		Bowatenna	GOV3
		Rantembe	GOV3

[a]When run the simulation load flow was stopped with 'Upper Kotmale PP with GOV3.' This could be eliminated by using GOV1
[b]% over shoot of power (of SapugaskandaPP) during starting of the generator (during the transient period) could be minimized with GOV3 than with GOV2

Fig. 3.32 Governor models used in PSCAD; **a** Steam_Gov_2 (GOV2) **b** Steam_Gov_3 (GOV3)

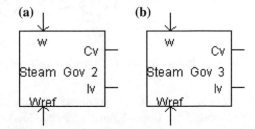

As per [11], the block diagram of 'Electro-hydraulic controls governing system' is appeared in Fig. 3.36. By comparing the block diagrams in Figs. 3.35 and 3.36, typical values for corresponding parameters of PSCAD control system were decided. These values are appeared in Table 3.17

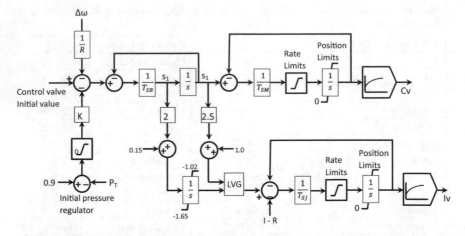

Fig. 3.33 Block diagram of control system of Steam_Gov_2 (Mechanical-hydraulic controls), PSCAD simulation model. Where, Cv—Control valve floor area (p.u.); Iv—Intercept valve floor area (p.u.); R—Permanent Droop (p.u.); T_{SJ}—IV servo time constant (s); T_{SM}—Gate servo time constant (s); T_{SR}—Speed relay lag time constant (s)

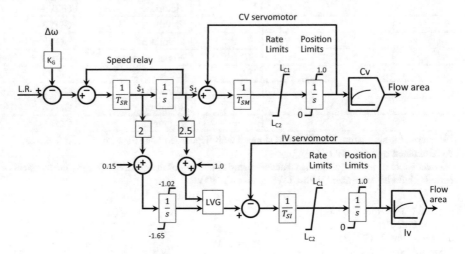

Fig. 3.34 Block diagram of 'Mechanical-hydraulic controls governing system' as per [11]

The hydro turbine governor used mostly is the Hydro_Gov_3 (GOV3) but only for modelling Upper-Kotmale-PP the Hydro_Gov_1 (GOV1) was used. Hydro Governor-1 and 3 as appeared in PSCAD are shown in Fig. 3.37a, b respectively.

The block diagrams of the control systems of hydro governor 'Hydro_Gov_1 (Mechanical-hydraulic controls (GOV1))' and 'Hydro_Gov_3 (Enhanced controls for load rejection studies (GOV3))' are appeared in Figs. 3.38 and 3.40 respectively.

Table 3.16 Typical values for the parameters in Mechanical-Hydraulic Controls (GE) (GOV2) model in PSCAD

Parameter as per [11]	Corresponding parameter in PSCAD	Typical value
K_G	1/R	20
T_{SR}	T_{SR} (Speed Relay time constant)	0.1 s
T_{SM}	T_{SM} (CV servo-motor time constant)	0.2 s
T_{SI}	T_{SJ} (IV servo-motor time constant)	0.2 s
L_{C1}	P_up	0.2 pu/s
L_{C2}	P_down	−0.5 pu/s
L_{I1}	I_up	0.2 pu/s
L_{I2}	I_down	−0.5 pu/s

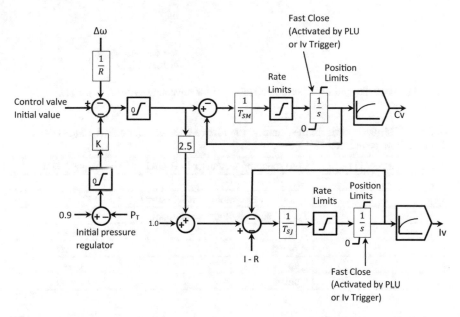

Fig. 3.35 Block diagram of control system of governor Electro-Hydraulic Controls (GOV3), as per PSCAD simulation model. Where, Cv—Control valve floor area (p.u.); Iv—Intercept Valve Flow Area (p.u.); R—Permanent Droop (p.u.); T_{SJ}—IV Servo Time Constant (p.u.); T_{SM}—Gate servo time constant [p.u.]; P_T—Steam pressure (constant)

As per [11], the block diagram of 'Electro-hydraulic controls governing system' is appeared in Fig. 3.39. By comparing the block diagrams in Figs. 3.38 and 3.39, typical values for corresponding parameters of PSCAD control system were decided. These values are appeared in Table 3.18 (Fig. 3.40).

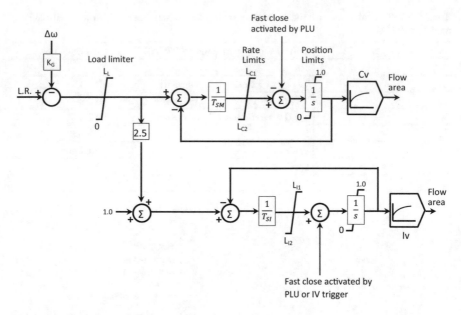

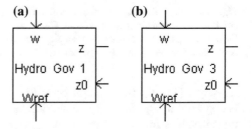

Fig. 3.36 Block diagram of 'Electro-hydraulic controls governing system' as per [11]

Table 3.17 Typical values for the parameters in Electro-Hydraulic Controls (GE) (GOV2) model of a steam turbine in PSCAD

Parameter as per [11]	Corresponding parameter in PSCAD	Typical value
K_G	1/R	20
T_{SM}	T_{SM} (CV servo-motor time constant)	0.1 s
T_{SI}	T_{SJ} (IV servo-motor time constant)	0.1 s
L_{C1}	P_up	0.1 pu/s
L_{C2}	P_down	−0.2 pu/s
L_{I1}	I_up	0.1 pu/s
L_{I2}	I_down	−0.2 pu/s

Fig. 3.37 Governor models used in PSCAD;
a Hydro_Gov_1 (GOV1)
b Hydro_Gov_3 (GOV3)

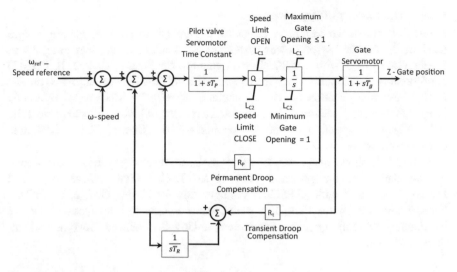

Fig. 3.38 Block diagram of the control system of hydro governor 'Hydro_Gov_1 (Mechanical-hydraulic controls (GOV1) as per PSCAD simulation model). Where, Q—servo Gain (p.u.); Rp—Permanent Droop (p.u.); Rt—Temporary Droop (p.u.); Tg—Main Servo time constant (s); Tp—Pilot valve and servo motor time constant (s); Tr—Reset or Dashpot time constant (s)

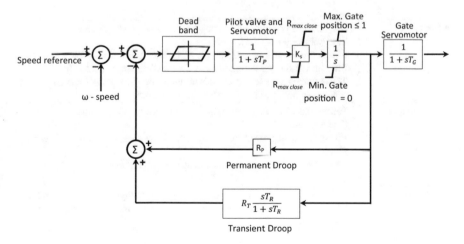

Fig. 3.39 Block diagram of governing system for hydraulic turbine as per [3]

With reference to [11], the block diagram shown in Fig. 3.39 and sample data in Table 3.19 provide values to configure the governor model 'Hydro_Gov_3 (Enhanced controls for load rejection studies (GOV3))' of PSCAD.

Values Used with PSCAD Window

Figure 3.41 shows the parameter values for thermal (steam) turbine governor types used in the simulation of power system of Sri Lanka: Steam_Governor_2 and Steam_Governor_3 with a. PSCAD configuration for Steam _Gov_2, b. PSCAD configuration for Steam _Gov_3, c. parameters of Gov2: Mechanical-Hydraulic (GE) Governor: section 1, d. parameters of Gov2: Mechanical-Hydraulic (GE) Governor: section 2, e. parameters of Gov3: Electro-Hydraulic (GE) Governor: section 1 and f. parameters of Gov3: Electro-Hydraulic (GE) Governor: section 2.

Figure 3.42 shows the parameter values for hydro turbine governor types used in the simulation of the power system of Sri Lanka: Hydro_Governor_1 and Hydro_Governor_3 with a. PSCAD configuration for Hydro_Gov_1, b. PSCAD configuration for Hydro_Gov_3, c. parameters common to both types, d. Gov_1 Mechanical-hydraulic governor parameters, e. Gov_3 Enhanced Governor parameters are given.

3.3   Control System of the Overall Power System

The control system of the overall power system is comprised with three modules. They are:

• LS Logic1	• Senses the frequency and voltage of the system (Biyagama Grid-substation) • Filters the frequency and voltage with 'Low pass, Chebyshev, Order 3' filter and 'Low pass, Butterworth, Order 2' filters respectively • These values are used to implement the under frequency Load Shedding scheme (without going for islanding operation)
• U_Frequency	• Senses the frequency and voltage of the system (Biyagama Grid-substation) • Filters the voltage and frequency with 'Low pass, Chebyshev, Order 3' filters • These values are used to implement the tripping action of generators during low frequencies (f < 47 Hz)
• Add_Ld	• To see the system response (frequency and voltage) after a sudden load addition in the power system (here it was performed at Biyagama grid sub-station)

3.3.1   LS Logic1 Control System Module

Two types of filters were used in measuring 'system frequency' and 'system voltage.' For the measurement of the system frequency, Low pass, Chebyshev,

Table 3.18 Typical values for the parameters in Mechanical-Hydraulic Controls (GE) (GOV1) model of a hydro turbine in PSCAD

Parameter as per [11]	Corresponding parameter in PSCAD	Typical value
T_P	T_P	0.05 s
K_S	Q	5.0
T_G	T_g	0.2 s
R_P	R_P	0.04
R_T	R_t	0.4
T_R	T_R	5.0 s
Max. gate position	Gmax	1
Min. gate position	Gmin	0
$R_{max\ open}$	MXGTOR	0.16 pu/s
$R_{max\ close}$	MXGTCR	0.16 pu/s

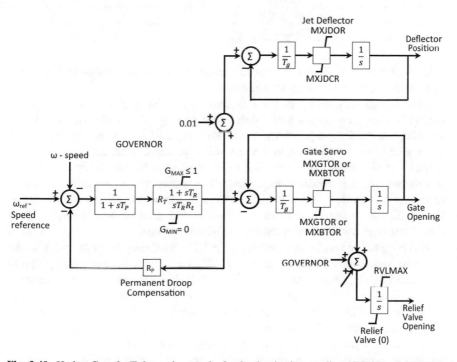

Fig. 3.40 Hydro_Gov_3 (Enhanced controls for load rejection studies (GOV3) of PSCAD). Where, G_{MAX}—Maximum Gate Position (p.u.); G_{min}—Minimum Gate Position (p.u.); MXBGCR —Maximum Gate Buffer Closing Rate (p.u./s); MXBGOR—Maximum Gate Buffer Opening Rate (p.u./s); MXGTCR—Maximum Gate Closing Rate (p.u./s); MXGTOR—Maximum Gate Opening Rate (p.u./s); MXJDCR—Maximum Jet Deflector Closing Rate (p.u./s); MXJDOR—Maximum Jet Deflector Opening Rate (p.u./s); RVLMAX—Maximum Relief Valve Position (p.u.); RVLCVR—Relief Valve Closing Rate (p.u./s); R_p—Permanent Droop (p.u.); R_t—Temporary Droop (p.u.); T_g—Main Servo Time Constant (s); T_p—Pilot valve and servo motor time constant (s); T_R—Reset or Dashpot Time Constant (s)

Table 3.19 Values used to configure the governor model 'Hydro_Gov_3 (enhanced controls for load rejection studies (GOV3))'

Parameter as per [11]	Corresponding parameter in PSCAD	Typical value
T_P	T_P	0.05 s
K_S	Q	5.0
T_G	T_g	0.2 s
R_P	R_P	0.04
R_T	R_t	0.4
T_R	T_R	5.0 s
Max. gate position	Gmax	1
Min. gate position	Gmin	0
$R_{max\ open}$	MXGTOR	0.16 pu/s
$R_{max\ close}$	MXGTCR	0.16 pu/s
	MXBGOR	
	MXBGCR	
	MXJDOR	
	MXJDCR	

order-3 filter was employed and for the measurement of the system voltage, Low pass, Butterworth, order-2 was employed.

Even though the Chebyshev has a ripple in its pass-band, it optimizes the roll-off time [47]. The main intention of the design is to suggest an under frequency load shedding scheme for the power system of Sri Lanka. So it is necessary to filter-out the required low frequencies to get the actual power system frequencies during normal and abnormal situations. The order of the Chebyshev filter was decided experimentally. It is required to maintain the system frequency at 50 Hz under steady state condition. Selection of the type of filters was done experimentally. Table 3.20 shows the 'Maximum overshoot during start of source' and 'Steady state frequency' obtained with different Chebyshev filters.

Referring to the results obtained in Table 3.20, the Chebyshev filter with 'Order = 3, Base frequency = 3 Hz, Ripple = 0.05 dB, Maximum overshoot during start of source =55.01 Hz, Steady state frequency = 50 Hz' was decided to use in detection of frequency, in modeling the control system of the power system. Figure 3.43 shows the PSCAD windows corresponding to the Chebyshev filter configuration.

Since the Butterworth filter optimizes the flatness of the passband [47], a 'Low pass, Order 2, Butterworth filter with base frequency 0.5 Hz' was used to filter-out the voltage signal for measuring purposes. Figure 3.44 shows the PSCAD windows for Butterworth filter configuration.

Figure 3.45 shows the Chebyshev and Butterworth filters used in measurement of System frequency and System voltage respectively, in the control system of the simulation model.

In-order to get the rate of change of frequency, it is possible to use the model 'derivative with a time constant' in PSCAD which is shown in Fig. 3.46. As per the PSCAD help itself says [37], 'This block is fraught with danger because of its

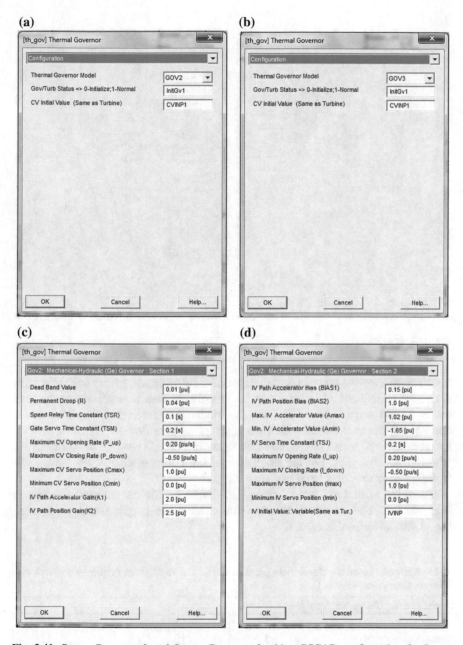

Fig. 3.41 Steam_Governor_2 and Steam_Governor_3 with **a** PSCAD configuration for Steam _Gov_2, **b** PSCAD configuration for Steam _Gov_3, **c** parameters of Gov2: Mechanical-Hydraulic (GE) Governor: section 1, **d** parameters of Gov2: Mechanical-Hydraulic (GE) Governor: section 2, **e** parameters of Gov3: Electro-Hydraulic (GE) Governor: section 1, **f** parameters of Gov3: Electro-Hydraulic (GE) Governor: section 2

(e)   **(f)**

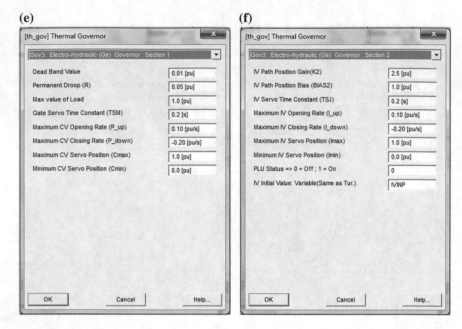

Fig. 3.41 (continued)

tendency to amplify noise. To minimize noise interference, particularly when the derivative time constant is large and the calculation step is small, it may be necessary to add a noise filter.' Further, the model time interval Δt can't be varied considering the situation to which we apply. This considers the value of the 'solution time step' which is the time step taken in PSCAD runtime to work out the whole simulation [30]. For the simulation of the power system of Sri Lanka, this 'solution time step' has been taken as 100 μs.

There for to find the rate of change of frequency (df/dt), the model shown in Fig. 3.47 has been used. It clearly demonstrates the output function given in Fig. 3.46b, [48]. In this model,

Table 3.20 Maximum overshoot during start of source and steady state frequency obtained with different Chebyshev filter settings

Details of the Chebyshev filter used					
Order	2	3	3	3	4
Base frequency (Hz)	3	3	3	3	3
Ripple (dB)	0.05	0.05	0.1	0.03	0.05
Maximum overshoot during start of source (Hz)	52.66	55.01	55.08	54.92	56.56
Steady state frequency (Hz)	49.75	50	50.01	$50.00 \leq f \geq$ 50.015	49.76

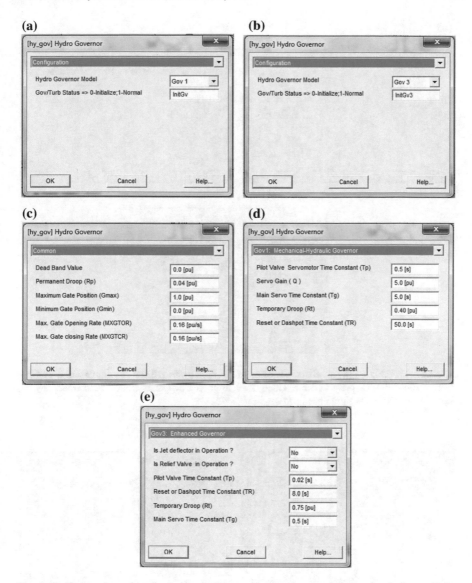

Fig. 3.42 Parameter values for hydro turbine governor types used in the simulation model a PSCAD configuration for Hydro_Gov_1, **b** PSCAD configuration for Hydro_Gov_3, **c** parameters common to both types, **d** Gov_1 Mechanical-hydraulic governor parameters, **e** Gov_3 Enhanced Governor parameters

Delay time, T	=0.01 s
Number of samples in delay T	=1
Time step of derivative, Δt	=0.01 s

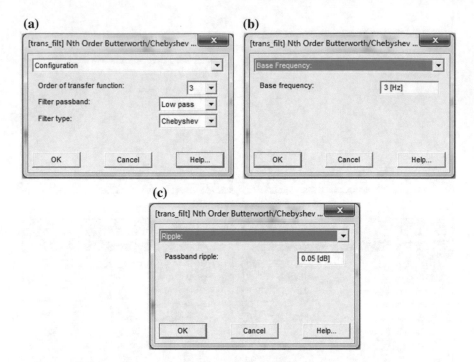

Fig. 3.43 PSCAD windows corresponding to the Chebyshev filter configuration a PSCAD configuration of Low pass, Order 3, Chebyshev filter, **b** Chebyshev filter's base frequency has been set to 3 Hz, **c** its ripple has been set to 0.05 dB

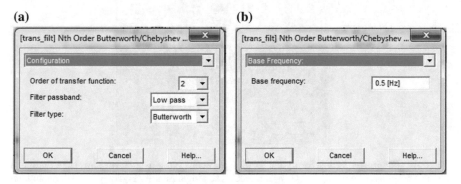

Fig. 3.44 PSCAD windows with **a** Butterworth filter configuration, **b** base frequency set to 0.5 Hz

With the available power generation, the power system can withstand a sudden generation deficit of 20 MW. At that instant the rate of change of frequency is $\approx$ −0.03 Hz/s. The relay settings for load shedding should be done only if the output signal y(t) < −0.03 Hz/s. A delay time that corresponds to relay operating time, circuit breaker operating time and an intentionally set delay time, should also be

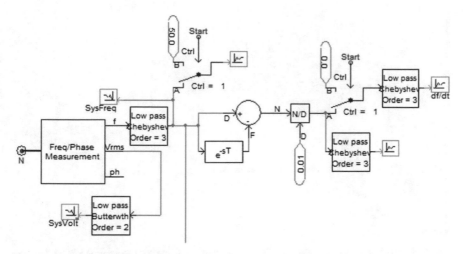

Fig. 3.45 The Chebyshev and Butterworth filters used in measurement of system frequency and system voltage respectively, in the control system of the simulation model

(a)

$$x(t) \longrightarrow \boxed{sT} \longrightarrow y(t)$$

(b)

$$y(t) = \frac{x(t) - x(t - \Delta t)}{\Delta t}$$

Fig. 3.46 Derivative function in PSCAD; **a** block diagram of the DERIV function, **b** the output as a function of the input

Fig. 3.47 PSCAD model used to calculate the df/dt of the power system

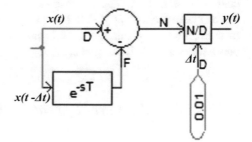

considered in sending a signal for the relay to operate. The PSCAD model 'sample and hold' is used to send the correct signal to the relay to get operated. This corresponds to different stages of the load shedding scheme. Figure 3.48 shows the control circuit of this operation.

Figure 3.49 demonstrates how the system responds to a sudden generation deficit of 23.18 MW (by tripping generator Wimalasurendra). The steady state frequency and voltage it achieves are 49.5 Hz and 211.81 kV respectively. With this disturbance it achieves a frequency degradation rate of −0.038 Hz/s. Victoria2

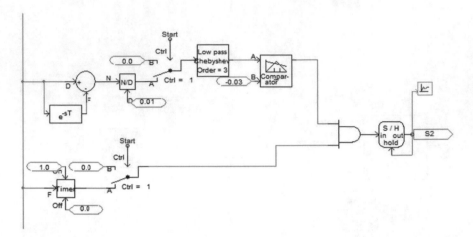

Fig. 3.48 The control circuit which senses the system frequency and to operate circuit breakers

which has been set as the isochronous governor reaches to its maximum output with this situation.

3.3.2 U_Frequency Control System Module

The same frequency measuring control techniques have been employed as discussed in Sect. 3.3.1. The purpose of the control system module is to trip off the thermal generators at 47 Hz after sensing the system frequency. Since the operation of thermal power plants are entirely not good for its performance and their operation, this action is taken place.

3.3.3 Add_Ld Control System Module

This control system module is to see the power system behavior during a sudden load addition (e.g. 5.75% of the total demand). The control logic has been done using sequencer components as shown in Fig. 3.50.

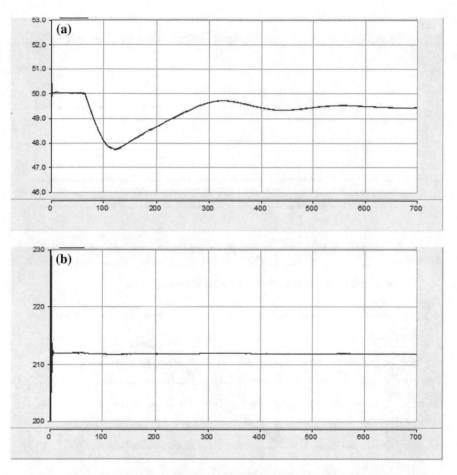

Fig. 3.49 System responses to a sudden generation deficit of 23.18 MW (by tripping generator Wimalasurendra); **a** Steady state frequency achieved = 49.5 Hz; **b** Steady state voltage achieved = 211.81 kV; **c** generation output of Victoria2 which has been set as the generator with isochronous governor; **d** rate of change of frequency in response to the disturbance—the minimum rate achieved = −0.038 Hz/s

3.4   Verifying the Simulation Model Performance

3.4.1   Steady State Operation

As reported in previous sections of this chapter, the power system simulation model was designed with PSCAD/EMTDC software. So it is very important to see its performance during normal and abnormal conditions of the power system. The load flow on the 13th March, 2013 during daytime peak demand has been simulated in the simulation model. It is very important to see how well it performs compared to the actual power system components in operation at the same instant.

(c)

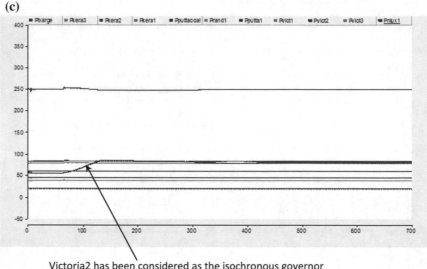

Victoria2 has been considered as the isochronous governor

(d)

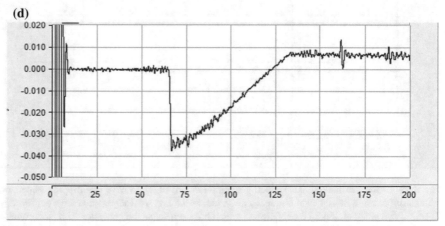

Fig. 3.49 (continued)

Corresponding actual generator outputs and simulation model's generator outputs of the load flow on the 13th March, 2013 during day time peak demand are tabulated in Table 3.21.

The voltage, frequency and some of the generator outputs are shown in the Fig. 3.51 under steady state operation. The program was run for 100 s. the lines of the power generation outputs corresponds to Coal power land (248.8 MW), Kerawlapitiya1 (82.25 MW), Kerawalapitiya3 (78.43 MW), Victoria1 and 3 (38.55 MW), Victoria2 (55.67 MW), Puttalam1 (47.58 MW), NewLaxapana1 (19.47 MW) and NewLaxapna2 (19.6 MW).

(a)

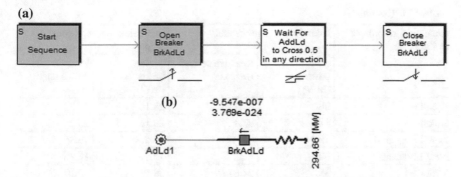

(b)

Fig. 3.50 A sudden load addition at the required instant, which was implemented using sequencer element; **a** series of sequencer elements used for the operation, **b** a resistive load was connected to the system through the breaker 'BrkAdLd'

Table 3.21 Actual generator outputs and simulation model's generator outputs corresponding to the load flow on the 13th March, 2013 during day time peak demand

Generators committed to the system	Actual generator outputs		Simulation model's generator outputs	
	Active power (MW)	Reactive power (MVAR)	Active power (MW)	Reactive power (MVAR)
Barge	60.5	35	59.7	27.62
KCCPgas	99	77	89.35	58.97
KCCPsteam	54	29.5	49.74	28.71
Sapugaskanda-1	25.5	8	17.69	16.45
Sapugaskanda-2	18	8	17.69	16.79
Sapugaskanda-3	12.5	12	35.37	29.68
Kerawalapitiya-1	90	50	82.23	10.68
Kerawalapitiya-2	90	50	82.43	43.54
Kerawalapitiya-3	90	50	78.41	31.93
KHD (Asia power)-1	17.5	2	16.95	26.12
KHD (Asia power)-2	17.5	2	16.95	26.12
Puttalam Coal	247	110.8	248.7	64.99
Upper kotmale	60	19.6	69.16	50.71
Heledanavi-1	40.15	5.9	47.58	3.4
Heledanavi-2	40.15	5.9	47.58	3.4
Kotmale-3	67	45	69.71	63.1
Polpitiya-1	32.3	17	28.03	21.89
Polpitiya-2	32	17	27.76	21.91
Canyon-1	0			
Canyon-2	0			

(continued)

Table 3.21 (continued)

Generators committed to the system	Actual generator outputs		Simulation model's generator outputs	
	Active power (MW)	Reactive power (MVAR)	Active power (MW)	Reactive power (MVAR)
Kukule-1	26	8.61	28.16	23.55
Kukule-2	26	8.61	28.16	23.55
New Laxapana-1	18.34	−0.421	19.6	16.86
New Laxapana-2	30	−0.278	19.6	16.86
Victoria-1	35	25.67	38.51	43.14
Victoria-2	50	26.65	56.22	41.27
Victoria-3	35	25.67	38.5	43.14
Ukuwela	20	10	20.71	12.44
Randenigala	48	20	44.49	32.64
Wimalasurendra	25	0	23.48	10.64
Laxapana-1	28.5	−3	24.92	15.39
Laxapana-2	22.5	−2	15.63	16.65
Samanalawewa-1	40	0.17	37.43	9
Samanalawewa-2	40	0.17	37.43	9
Bowatenna	11	9.23	8.78	5.36
Rantembe-1	15	1.74	24.1	14.6
Embilipitiya	44.3	30	38.48	24.38
Total	1607.4	705.52	1589.23	904.48

3.4.2 Generator Tripping/Sudden Generation Deficit Situation

It is necessary to see whether the simulation model responds to some emergency situation same as the power system faces that. If the actual results are very much closer to the simulation model's results, then it can be suggested that the model developed could be used to see the power system behavior under different scenarios. Norochcholai Coal Power plant tripping, which occurred at a time closer to the considered system loading (day peak time on 13th May, 2013) that lead for a total black out was simulated and observed the system performance.

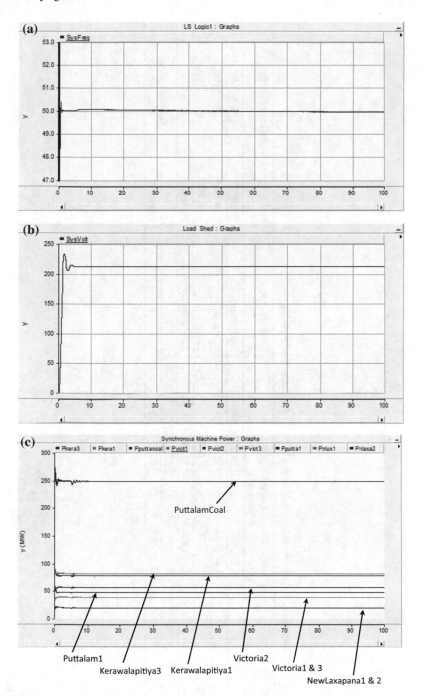

Fig. 3.51 Power system performance during steady state; **a** power system frequency, **b** power system (transmission) voltage, **c** some of the generator power outputs

Table 3.22 Removed generator outputs and loads from the PSCAD simulation model designed

Generators removed	Power delivered at the instant of the removal from the system		Loads removed	Power drawn at the instant of the removal from the system	
Victoria 1	37.35 MW	34.37 MVA	Biyagama	70 MW	30 MVA
Victoria 2	54.71 MW	32.64 MVA	Kotugoda	71.2 MW	25.4 MVA
Kerawalapitiya1	80.35 MW	2.6 MVA	Pannala	33.8 MW	21.8 MVA
KHD 1	16.66 MW	12.49 MVA	Aniyakanda	30 MW	18 MVA
KHD 2	16.66 MW	12.49 MVA			
TOTAL	205.73 MW	94.59 MVA		205 MW	95.2 MVA

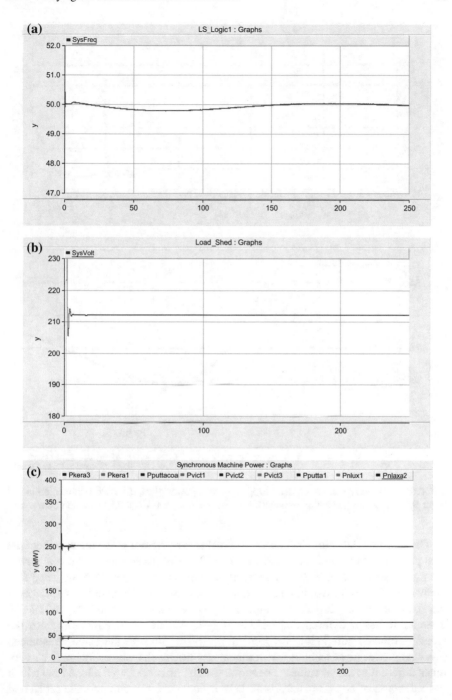

Fig. 3.52 After removal of some selected generator sets and loads (given in Table 3.22) from the simulated load flow, **a** system frequency; **b** system voltage; **c** power outputs of some generators

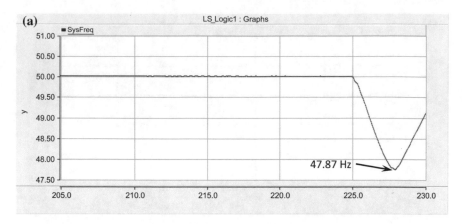

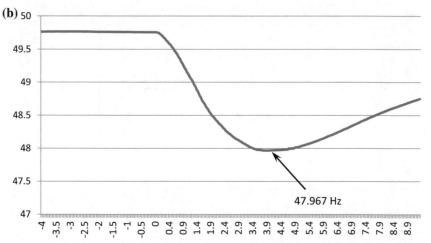

Fig. 3.53 Frequency profiles during a tripping of coal power plant; **a** with the simulation model (with some load and generation removal), **b** using actual system frequency variation data

The received data corresponds to an actual power system performance where the coal plant tripping occurs at a load flow of 1434 MW. Even though the coal power plant generates 300 MW, for its auxiliary services it consumes 25 MW. Therefore the balance 275 MW would be consumed by the system. This amounts 19.18% of the total demand. The already simulated model has a demand of 1637 MW. Therefor it has a difference of 203 MW ($\approx$12.4% demand reduction) from the already modeled simulation program. If we try to simulate the same phenomena (i.e. coal power plant tripping) with this model the results we receive may be very much different to actual values. Therefor a simulation model with a load flow which has a demand of approximately 1437 MW is considered to compare the actual and simulation model results. To get the 1437 MW load flow, several power plants and loads were removed from the original simulation. The removed generators and loads with their capacities are given in Table 3.22. Further, the coal power plant generates 250 MW with this load flow. It amounts 17.4% from the total demand.

With the above amendment, frequency, voltage and other selected generator outputs are shown in Fig. 3.52, under steady state condition.

Figure 3.53a shows the frequency profile when implement the load flow on 13th March, 2013 during peak load demand, with the amendments given in Table 3.22, while (b) shows the actual frequency profile during the generator tripping causing a generation deficit of 275 MW for the system. According to this comparison the simulation model reaches a minimum frequency of 47.87 Hz (a) where the actual system reaches a minimum frequency of 47.967 Hz (b).

So with the results shown in Fig. 3.53, it can be considered that the simulation model can give results much closer to the actual power system performances.

Chapter 4
Designing the Load Shedding Scheme

4.1 Overview of the Power System of Sri Lanka

Power System of Sri Lanka has a daily maximum power demand of around 2100 MW. This always takes place as the night peak of the load curve of the country. Out of the total electricity generation, about 65% is based on thermal and the majority of the balance is from hydro power plants. About 200 MW of the total generation is received from embedded generation which comprises with hydro and wind power [49]. 900 MW 'Lakwijaya' power station has the highest generation capacity in Sri Lanka [50]. It comprises with three 300 MW generators that operate on coal and possess around 25% of the total installed capacity of Sri Lanka. Therefore it performs a very important role in economic aspects, since its unit cost of generation is much less compared to that of other thermal power plants. Hence if a failure occurs and if it happened to isolate it from the national grid, there will be a considerable impact on the power system stability that may lead to a brown out or a black-out situation [51]. On the other hand, with the increased penetration of intermittent renewable energy, power systems encounter a high possibility of uncertainty and variability causing generation and load imbalances. Thus, a well-designed Under Frequency Load Shedding Scheme is critical to safe-guard the system during extreme situations to avoid complete system black-outs while maintaining its stability.

4.2 Identification of Parameters

In designing the load shedding scheme, parameters such as

- Power system regulations and practice of Sri Lanka
- Identifying settling frequency
- Number of steps in the load shedding scheme

© Springer Nature Singapore Pte Ltd. 2018
T. Bambaravanage et al., *Modeling, Simulation, and Control of a Medium-Scale Power System*, Power Systems, https://doi.org/10.1007/978-981-10-4910-1_4

- First step of load shedding scheme
- Identifying when to implement
- Delay time,

Should receive equal attention. Effect of these parameters in an under frequency load shedding scheme influence the power system stability very much.

4.2.1 Power System Regulations and Practice of Sri Lanka

According to the power system regulations and practice of Sri Lanka [52], it has been identified that:

- Voltages at the live bus-bars of CEB network:

  $132 \pm 10\%$ kV
  $220 \pm 5\%$ kV

- System frequency:

  Normal operating range: $50 \pm 1\%$ Hz

- According to the current practice of CEB,

  Normal operating range: $50 \pm 4\%$ Hz
  Short term variations: −6 to +5% up to 3 s.

Table 4.1 gives the Load Shedding Scheme that is being implemented in Sri Lanka (as in 2015).

From Step 1 to Step 5, the Load shedding action is sensitive to the system frequency only. Once the relay senses the frequency, if it meets the requirements given, after 100 ms (as an example lets refer to Step 1), it sends a signal to the

Table 4.1 Present Ceylon Electricity Board (CEB) load shedding scheme

Frequency state	Time delay and magnitude of LS
f ≥ 48.75 Hz	No load shed
f < 48.75 Hz	Step 1: 6.5% of load 100 ms time delay
f < 48.5 Hz	Step 2: 6.5% of load 500 ms time delay
f < 48.25 Hz	Step 3: 12% of load 500 ms time delay
f < 48 Hz	Step 4: 9% + 3.5%* = 12.5% of load 500 ms time delay
f < 47.5 Hz	Step 5: 3% + 4.5%* = 7.5% of load Instantaneous
f < 49.0 Hz and −0.85 < df/dt	13% + 3.5%* + 4.5%* = 21% of load, 100 ms time delay

circuit breaker to get operated. The actual time delay of this process is much more than the time delay it states here. When consider the last logic state: f < 49.0 Hz and −0.85 < df/dt, it is more adaptive to the system states.

4.2.2 Identifying Settling Frequency

Rotating machineries are designed for their optimum performance at a specific frequency—here this research considers that as 50 Hz operation. For the safe and effective operation of generators, frequency should not go below few percent of the rated frequency. Continuous operation of steam turbines should be restricted to frequencies above 0.99 p.u. (for 50 Hz base it is 49.5 Hz). Operation below 49.5 Hz should be limited to very short durations. Hence it is very important to co-ordinate the settling frequency of the PS and the relay tripping frequency in use to control generator shut-down [3, 31, 35].

In order to improve the quality of power, it has been considered that the power system of Sri Lanka should have a rated frequency of 50 Hz with below tolerances under the given conditions. The normal operating range has been considered as the settling frequency range during steady state operation, after a disturbance.

- System Frequency—Normal operating range: 50 Hz ± 1%
- System Frequency—During an emergency/transient: 50 Hz ± 4%
  Short term variations: −6 to +5% up to 3 s.

4.2.3 Deciding the Number of Steps in the Load Shedding Scheme

It is very important to decide a correct number of Load Shedding steps as it may cause over shedding for a small disturbance or less severe over-load. As a solution it can be considered to increase the number of steps and divide the load among them. Proper co-ordination among steps is equally important.

- Currently Power System of Sri Lanka comprises 6 no. of stages in its Load Shedding Scheme.
- In [31] and [53] it has been mentioned that '3–5 steps' is the most adequate selection based on experience.
- In [28] a comparison of several Under Frequency Load Shedding Schemes has been carried out which were with different number of stages (e.g. 6, 8 and more than 8 number of stages).

Therefore for this proposed Load Shedding Scheme, it has been considered the step size as six.

4.2.4 First Step of Load Shedding Scheme

With reference to the present load shedding scheme of Sri Lanka, it initiates its load shedding action at a frequency of 48.74 Hz. This is a value that is far below the system normal operating range. Many utilities that operate with a rated frequency of 60 Hz set the first step of load shedding at 59.5 Hz [53]. That is,

The 1st occurs at $59.5/60 = 0.99$ p.u.
With 50 Hz base frequency,

The first step $= 0.99 \times 50 = 49.5$Hz
But, Power System of Sri Lanka's normal operating frequency (f) range is, $50.5 > f > 49.5$ Hz. Hence, 49.4 Hz is a better option for the first step of the load shedding scheme. Further, initial under-frequency load shedding relay settings are typically 59.3 Hz in the United State systems [54]. For such situations,

The 1st occurs at $59.3/60 = 0.988$ p.u.
With 50 Hz base frequency,

The first step $= 0.988 \times 50 = 49.4$ Hz
In [53] it has been justified for a power system operating at 60 Hz can have a load shedding scheme with first step as 59.5 Hz. The same arguments referring to a system frequency of 50 Hz can be considered as below.

- All of the larger turbine generators on the system are not rated for continuous generation below 0.99 p.u. (for 60 Hz it is 59.5 Hz) depending on the manufacturer. Thus setting the initial load shedding frequency at a relatively high value, such as 49.4 Hz (for 50 Hz power system), also tends to limit the maximum frequency deviation.
- A load shedding program starting at 49.4 Hz can be more effective in minimizing the depth of the under frequency response for a heavy over load, than would a similar program which has a lower 1st shedding frequency.
- Hence for a 50 Hz power system, it has been argued and decided to consider the first step of the under frequency load shedding scheme as 49.4 Hz (instead of considering 48.75 Hz).

4.2.5 Identifying When to Implement Load Shedding Based on Rate of Change of Frequency (ROCOF)

According to the present CEB scheme:

If, ROCOF < -0.85 Hz/s and
System Frequency $= 49$ Hz, a 20% of the load is to be shed.

But, practical as well as simulation results show that the maximum ROCOF occurs at an instant after about 1.4–1.6 s from the moment of generation deficiency occurs. If we assume that the disturbance occurs at a system frequency of 49.5 Hz and this disturbance tends to cause a ROCOF = −0.85 Hz/s, simulation results show that the time consumed to achieve a frequency of 49 Hz is much less than 1.4 s. So the response corresponding to the conditions "f < 49.0 Hz and −0.85 < df/dt" of the CEB scheme occurs with much delay. Since this may lead to initiate initial stages of the current CEB LSS, there is a high chance of shedding an excessive amount of load. Figure 4.1a shows time taken to reach the maximum df/dt referring to a coal power plant tripping discussed in Sect. 3.4.2, with actual data. The plot in black is the df/dt with actual data while the smooth line in blue is the trend line. The minimum df/dt value achieved is—0.75 Hz/s. The same scenario simulated in the PSCAD simulation model takes approximately 1.6 s to reach the minimum df/dt. This is shown in Fig. 4.1b.

4.2.6 Delay Time

The delay time (DT) may vary from one load shedding scheme to another. But according to [31], the total amount of time necessary to clear the load can be considered as,

$$TD = R_{OT} + T_{Intentional} + B_{OT} \tag{4.1}$$

where,

TD   total clearing time
R_{OT} relay operating time
$T_{Intentional}$ intentional time delay (as a safety margin)
B_{OT} circuit breaker operating time

Power System of Sri Lanka uses a $T_{Intentional}$ of 0.1 s for its 1st step and 0.5 s for 2nd–4th steps. The condition "f < 49.0 Hz and −0.85 < df/dt", which is adaptive, operates with a $T_{Intentional}$ of 0.1 s. Its 5th step gets initiated with $T_{Intentional} = 0$ s.

- Chin-Chyr Huang et al. suggests a TD of 0.1 s for the under frequency load shedding simulation, but states it is not necessary to have the same time interval for each step [53].
- Quamrul Ahsan et al. used a time delay of 7.5 cycles ≃ 0.15 s (with a 50 Hz system) to implement 2nd and 3rd steps in their under frequency load shedding simulation program [8].

(a)

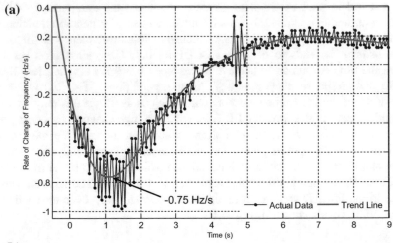

(b)

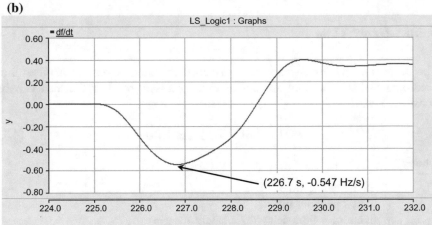

Fig. 4.1 Time taken to reach the minimum df/dt due to a coal power plant tripping; **a** actual power system performance (275 MW generator tripping); **b** simulation model results (250 MW generator tripping occurs at 225 s)

Further, referring to Eq. (4.1) and considering the following approximated values

$$TD = R_{OT} + T_{Intentional} + B_{OT}$$
$$\simeq 0.2\,s$$

where, $R_{OT} \simeq 60$ ms, $B_{OT} \simeq 83$ ms [35, 55] and $T_{Intentional} \simeq 60$ ms (three nos. cycles). Hence, for these proposed load shedding scheme simulations, a TD of 0.2 s was considered.

Since it is evident that Relays and circuit breakers take few milliseconds to get operated during a faulty situation, that time should also be included in the simulation program of the Present CEB load shedding scheme. The flow chart shown in Fig. 4.2 demonstrates the actual load shedding scheme considering the time factor ($R_{OT} + B_{OT} \simeq 140$ ms) involved in each load shedding stage. Table 4.2 shows the calculation of actual delay time involved in each load shedding action.

Fig. 4.2 Flow chart of the CEB load shedding scheme considering time delays for relay and circuit breaker operation ($R_{TO} + B_{OT} = 140$ ms) involved in each load shedding stage

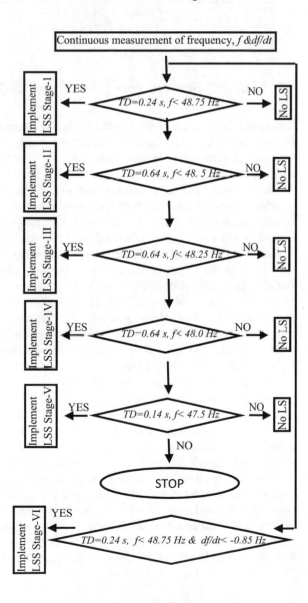

Table 4.2 Time delays corresponding to each load shedding action in the simulated present CEB load shedding scheme

Frequency state	Actual delay time considering the relay and circuit breaker operation
f ≥ 48.75 Hz	No load shed
f < 48.75 Hz	Step 1: 100 ms + 140 ms = 0.24 s
f < 48.5 Hz	Step 2: 500 ms + 140 ms = 0.64 s
f < 48.25 Hz	Step 3: 500 ms + 140 ms = 0.64 s
f < 48 Hz	Step 4: 500 ms + 140 ms = 0.64 s
f < 47.5 Hz	Step 5: 140 ms = 0.14 s
f < 49.0 Hz and −0.85 > df/dt	100 ms + 140 ms = 0.24 s

4.3 Ahsans' Scheme as a Pilot Model [56]

Bangladesh is a country located in the south Asian region as Sri Lanka. It is a small PS operates with 50 Hz, having an installed capacity of around 5200 MW, and an annual peak demand of around 4300 MW. As developing countries in the same region, both these countries may experience equal types of power system instability situations which lead for catastrophic events. Therefore it was decided to see the stability performance of the PS of Sri Lanka when the very same LS scheme, which is demonstrated in the Fig. 4.3 and Table 4.3, is applied. Hence it was carefully studied in order to simulate in the PS model of Sri Lanka.

According to Ahsan et al., their aim is to bring the system frequency up to 49.1 Hz with the under frequency load shedding scheme (with reference to the Bangladesh power system). This was decided on the capability of the power system to sustain without violating its stability conditions during a generation deficiency maximum of 7%. From there on wards, with the spin reserve available the system could be brought back to normal operating condition. The rate of change of frequency at the instant of 7% generation deficiency was 0.2 Hz/s. Hence f_{TH} = 49.1 Hz and m_0 = 0.2 Hz/s.

To implement the Ahsans' load shedding scheme in Sri Lanka power system, four islands were identified in the national grid [65, 66]. These islands have their own power generation, but it may not sufficient to meet the total demand. Therefor under normal operating conditions (operating as a part of the national grid; i.e. islanding operation is not implemented) these areas have the chance of receiving power from the national grid. This can be demonstrated as in Fig. 4.4.

The same load shedding scheme (Ahsans') has been applied to power system of Sri Lanka, referring to Sect. 4.1. The values f_{TH} and m_0 were considered as f_{TH} = 49.4 Hz, and m_0 = −0.03 Hz/s. When implement a sudden generation outage of 15% of the demand, how the frequency varies with time was observed. The simulation results showed that the time taken to change the frequency from 49.4 to 49.3 Hz is much less than 0.14 s, which is the approximate time taken for operation of

Fig. 4.3 Flow chart illustrating different steps of the technique introduced by Ahsan et al.

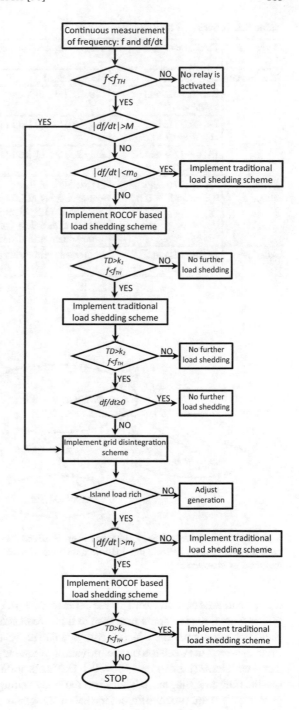

Table 4.3 Ahsans' LS Scheme

Types of LS	Frequ. state	Time delay and magnitude of LS
Traditional	f ≥ 49.1 Hz	No load shed
	f < 49.1 Hz and |df/dt| < 0.2	Step 1: 15% of load
	f < 49.0 Hz and |df/dt| < 0.2	Step 2: 15% of load (after a delay of 7.5 cycles)
	f < 48.9 Hz and |df/dt| < 0.2	Step 3: 15% of load (after a delay of 7.5 cycles)
Semi adaptive	f < 49.5 Hz and 0.2 < |df/dt| < 0.6	Step4: 50% of load • after a delay of 20 cycles if f < 49.1 Hz, implement traditional LS scheme • Next, after a delay of 20 cycles if f < 49.5 Hz, implement grid disintegration scheme
Adaptive	f < 49.5 Hz and 0.6 < |df/dt|	Implement grid disintegration scheme

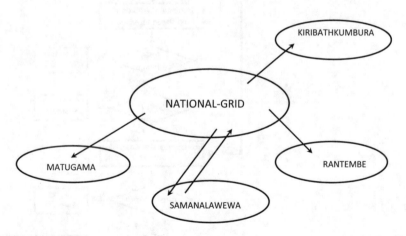

Fig. 4.4 Power system of Sri Lanka has been considered as a group of small islands and the national grid. *Arrows* indicate the possible directions of power flow—either from national grid to an island or vice versa

a relay and a circuit breaker (please refer to Sect. 4.1). Further it takes time for the power system to give some response to this circuit breaker operation. That is since it requires a minimum time which is equivalent to the relay and circuit breaker operating time, to take some decision on the prevailing frequency state, frequency gap between step 1 and step 2 (i.e. 49.4 − 49.3 = 0.1 Hz) is not sufficient (please refer to Fig. 4.5 to see the time line diagram). Therefore the relay settings corresponding to Sri Lanka power system are much different to that of Bangladesh power system.

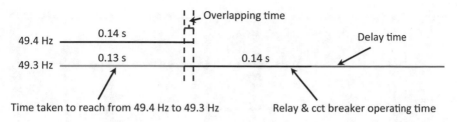

Fig. 4.5 Time line diagram that demonstrates the relay and circuit breaker operating times which get overlapped

Let's consider a situation where a generation deficit of 243.08 MW (by a forced outage of Kerawalapitiya 1, 2, and 3 generators with generations of 82.23, 82.43, 78.42 MW respectively).

According to the time line diagram, at 49.4 Hz relay and circuit breaker activation (R&B1) occur. They take 0.14 s to finish its operation. That means the power system initiates responding to this circuit breaker operation after 0.14 s duration. But before finishing the operation of R&B1, the relay and circuit breaker activation corresponding to 49.3 Hz get started. According to the simulation, time taken for the power system to reach from 49.4 to 49.3 Hz is 0.13 s. hence, there is an overlapping time of 0.01 s. therefore this may cause excessive load shedding. In turn the power system may become unstable due to over frequency situations. Therefore the suggested relay settings (49.4 Hz, 49.3 Hz etc.) corresponding to stage 1, 2, etc. may not address the under frequency situations successfully.

Figure 4.6 demonstrates the system frequency variation during a forced generation tripping of 243.08 MW (14.85% < 15%) in the simulation model. To address this generation deficit, two stages of the load shedding scheme got implemented.

As there are draw backs with the Ahsans' model when implement for the power system of Sri Lanka, it is very important to design a load shedding scheme considering different factors that are specific to Sri Lankan power system.

4.4 Proposed Methodology

4.4.1 Load Shedding Scheme-I (Based on Prevailing Facilities Available with the CEB)

54% of the prevailing load is involved in the load shedding scheme. It is comprised with 6 nos. of shedding stages that get operated based on system frequency and rate of change of frequency of the power system. Figure 4.7 demonstrates a flow chart of the proposed load shedding scheme-I. Table 4.4 briefs the suggested load shedding scheme-I. Since the system frequency goes beyond or below the rated

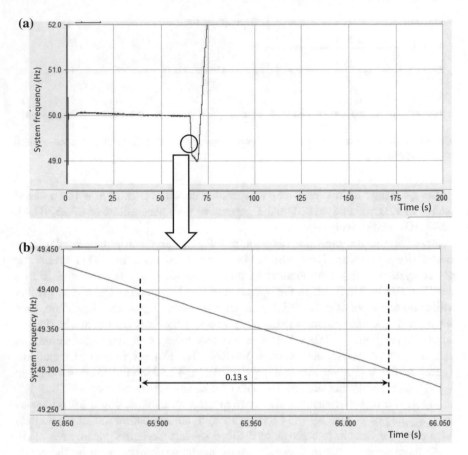

Fig. 4.6 a System frequency variation during a forced generation tripping of 243.08 MW (14.85%), **b** time taken to change the frequency from 49.4 to 49.3 Hz has been observed as 0.13 s

conditions given in Table 4.4, a power system instability situation may be possible, the system frequency and its 1st derivative are measured all throughout for load shedding actions to take place.

When actual system frequency variation against time as well as simulation results are analyzed, it is evident that usually, the power system reaches its df/dt a maximum after about 1.4–1.6 s from the instant of the occurrence of the disturbance (Fig. 4.1). To initiate the load shedding scheme, the system frequency should be dropped to a frequency of 49.4 Hz. From 1–2 to 3–4 load shedding stages, 6 and 8% of the prevailing load is shed respectively, based on the system frequency. At the 5th stage, the load shedding scheme is more adaptive to the system behavior as it gets implemented based on the rate of change of frequency of the power system, in which a 10% load is scheduled to be shed at −0.675 Hz/s. The 6th stage is once again based on a combination of system frequency as well as ROCOF with a load size of 10%. The 6th or the last stage gets implemented when the power system

Fig. 4.7 Flow chart demonstrating the load shedding scheme-I, that is comprised with 6 nos. of stages

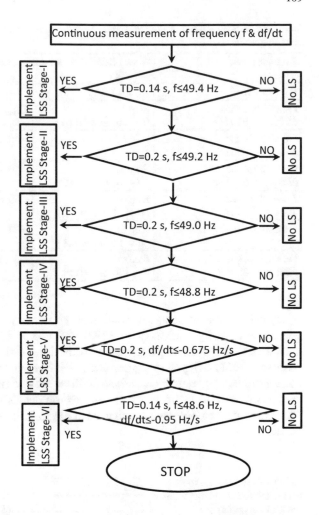

satisfies both the conditions df/dt $\leq$ −0.95 Hz/s and system frequency (SF) = 48.6 Hz. This amounts to 21% of the total load. With the implementation of the 6th stage, a 59% of the total prevailing load is scheduled to be shed. It is possible to decide which loads are to be shed, considering such factors as their importance to communities (e.g. hospitals), contribution to national economy (e.g. industrial zones) etc. It is very important to limit the system frequency reducing below 0.95 p.u. (For power systems with rated frequencies 60 and 50 Hz, the corresponding minimum safe frequencies are 57 and 47.5 Hz respectively). It may lead the power system unstable further due to tripping of steam turbine and gas turbine power plants [3]. This can badly affect the performance of induction motors in operation, which in turn gives negative effects on the national economy as well. For a power system mainly based on thermal-plants, it may cause a system brown-out or even a black-out.

Table 4.4 Proposed load shedding scheme-I with time delays corresponding to each load shedding action

Frequency state	Actual delay time considering the relay and circuit breaker operation	
f ≥ 49.4 Hz	No load shed	
f < 49.4 Hz and df/dt = −0.03 Hz/s	Step 1: 0 + 140 ms = 0.14 s	$T_{intentional}$ $+T_{BO} + T_{RO} = 0 + 80 + 60$ ms
f < 49.2 Hz and df/dt = −0.03 Hz/s		=60 + 80 + 60 ms
f < 49.0 Hz and df/dt = −0.03 Hz/s	Step 3: 60 + 140 ms = 0.2 s	=60 + 80 + 60 ms
f < 48.8 Hz and df/dt = −0.03 Hz/s	Step 4: 60 + 140 ms = 0.2 s	=60 + 80 + 60 ms
f < df/dt < −0.675 Hz/s	Step 5: 60 + 140 ms = 0.2 s	=60 + 80 + 60 ms
f < 48.6 Hz and df/dt < −0.95 Hz/s	Step 6: 0 + 140 ms = 0.14 s	=0 + 80 + 60 ms

By doing a forced outage of some selected power plants, addition of that has a generation capacity equivalent to 829.6 MW ($\approx$3 × 275 MW = 3 × coal power generation), the system performance with the Load Shedding Scheme-I was observed. Table 4.5 presents generators involved in the forced outage and the corresponding generation capacities of those power plants at that instant. The system was able to maintain its stability whose details are presented in Sect. 5.1. This was achieved by paying equal attention to both active and reactive power balance.

Table 4.5 Generators took part in the forced outage with their corresponding generation capacities (total capacity outage = 829.6 MW)

Generator	Capacity (MW)
Barge	59.69
KCCPgas	89.12
KCCPsteam	49.65
Puttalam 1	47.37
Coal	248.75
Kukule1	28.1
Kukule2	28.1
Kerawalapitiya1	81.82
Kerawalapitiya2	81.82
Kerawalapitiya3	77.93
Embilipitiya	37.24

4.4.2 Load Shedding Scheme-II (Based on Disintegration of the Power System)

It has been considered a load of 40% of the total demand of the power system as 'the uninterrupted power supply load (critical load segment)'. The Load Shedding Scheme is comprised with 2 nos. of stages, named Phase-I and Phase-II. During Phase-I, approximately 30% of the prevailing load is shed to maintain the stability. Disintegration of the power system is occurred under the Phase-II. If the power system cannot be brought to a stable operating condition with Phase-I of the Load Shedding Scheme, then the islanding operation is done. The Phase-I gets initiated if the system frequency reaches 48.6 Hz. It can be proved that if load shedding stages get implemented above 48.6 Hz and if the amount of load shed is sufficient to make the system stable, then the system frequency can be maintained above 47 Hz (closer to 47.5 Hz), which is the safe operating frequency for steam power plants. As the disintegration of the power system is done at 48.6 Hz, the National Grid's as well as the islands' stable operation is much assured.

To initiate the Load Shedding Scheme, the system frequency should be dropped to 49.4 Hz. Table 4.6 and Fig. 4.8 demonstrates the Phase-I of the Load Shedding Scheme-II: from 1, 2, 3 and 4 Load Shedding stages, 6, 6, 8 and 10% of the prevailing load is shed respectively, based on the system frequency degradation. This is approximately comprised of 30% of the total (prevailing) system load. The 5th stage is for the disintegration of the power system. Four possible locations in the power system of Sri Lanka were identified as load centres that can be operated as islands.

Boundaries of those locations suitable for Islanding [66] are:

- Island Matugama: (the load fed from Pannipitiya grid—Loads connected to Panadura, Horana, Matugama and Ambalangoda Bus-bars)
- Island Embilipitiya: (the load fed from Balangoda grid—Loads connected to Ratnapura, Embilipitiya, Hambantota, Deniyaya, Galle, Beliatta and Matara Bus-bars)

Table 4.6 Phase-I of proposed LSS-II with time delays corresponding to each LS action

Frequency state	Actual delay time considering the relay and circuit breaker operation	
$f \geq 49.4$ Hz	No load shed	
$f < 49.4$ Hz and $df/dt = -0.03$ Hz/s	Step 1: 0 + 140 ms = 0.14 s	$T_{intentional}$ $+T_{BO} + T_{RO} = 0 + 80 + 60$ ms
$f < 49.2$ Hz and $df/dt = -0.03$ Hz/s	Step 2: 60 + 140 ms = 0.2 s	$= 60 + 80 + 60$ ms
$f < 49.0$ Hz and $df/dt = -0.03$ Hz/s	Step 3: 60 + 140 ms = 0.2 s	$= 60 + 80 + 60$ ms
$f < 48.8$ Hz and $df/dt = -0.03$ Hz/s	Step 4: 60 + 140 ms = 0.2 s	$= 60 + 80 + 60$ ms

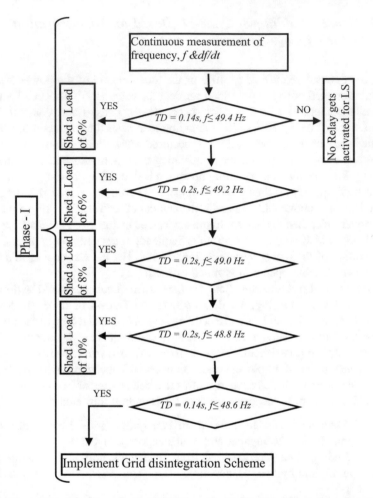

Fig. 4.8 Flow chart of the Phase-I of load shedding scheme-II which initiates Islanding at system frequency = 48.6 Hz

- Island Rantembe: (the load fed from Rantembe grid—Loads connected to Nuwara Eliya, Badulla, Inginiyagala and Ampara Bus-bars)
- Island Kiribathkumbura: (the load fed from Anuradhapura and New-Anuradhepura grids—Loads connected to Vavuniya, Kilinochchi, Trincomalee, Habarana, Valaichchenai, Ukuwela, Bowatenna and Kiribathkumbura Bus-bars)

By doing a forced outage of selected power plants, addition of them has a generation capacity equivalent to 495.14 MW, the system performance was observed. Phase-II of the load shedding scheme-II (Islanding operation) got implemented. Table 4.7 presents generators involved in the forced outage and the corresponding generation capacities of those power plants at that instant.

Table 4.7 Generators took part in the forced outage with their corresponding generation capacities (total capacity outage = 495.14 MW)

Generator	Capacity (MW)
Barge	59.69
KCCPgas	89.36
KCCPsteam	49.74
Puttalam 1	47.58
Coal	248.77
Total	495.14

The Load Shedding Scheme has been developed in such a way a 40% of the current demand in the National Grid should be given priority in catering electricity, if it is required to disintegrate the power system. So with regard to the load-flow concerned,

$$\text{Power demand} = 1637 \, \text{MW}$$

$$40\% \text{ of the demand} = 654.8 \, \text{MW}$$

Under a situation where the prioritized load is 40% of the demand, a demand (or a load) of 660.82 MW, which is approximately equal to 654.8 MW, has been retained, and the rest (or the balance) was tripped off.

Special attention should be paid to make the active power generation and consumption balance in the National Grid after the disintegration process (in a situation where 40% of the total electricity demand is catered).

Total generation of the national grid before the disturbance	=1589.23 MW
Forced outage (generation tripping to make islanding operation)	=495.14 MW
Generation in Islands	=56.34 + 28.68 + 109.38 + 24.17 =218.57 MW
Effective load connected to the national grid (i.e. 40% of the total demand)	=654.8 MW
Excess generation	={1589.23 − (495.14 + 218.57)} − 654.8 = (1589.23 − 713.71) − 654.8 = 220.72 MW

If 220.72 MW generation is tripped expecting a 100% stable operation, the outcome will be a rise in frequency beyond 52 Hz (exceeds 70 Hz). Increase of frequency indicates a generation high environment. Generation was reduced according to the following equation.

Required generation to be tripped = $G - (\alpha D)$
Simulation results showed that when $0.925 > \alpha > 0.903$, system becomes stable.

Table 4.8 Generators which were tripped off to balance the power generation and demand of the National Grid

Generator	Power generation (MW)	Purpose of operation
KHD2	16.95	TRIP to balance demand and load
Kerawalapitiya1	82.23	TRIP to balance demand and load
Kerawalapitiya2	82.43	TRIP to balance demand and load
Kerawalapitiya3	78.41	TRIP to balance demand and load
Total	260.0	

Islanding with $\alpha = 0.925$:

Out of the generators in the national grid, KHD2, Kerawalapitiya1, Kerawalapitiya2 and Kerawalapitiya3 were tripped off to balance the generation and demand at that instant. The generation capacities of corresponding generators are appeared in the Table 4.8.

$$\text{Required generation to be tripped} = G - (\alpha D)$$
$$= 867.27 - (0.925 \times 654.53)$$
$$= 261.83$$

As discussed, a disturbance was imposed on the power system with a forced generator capacity outage of 495.14 MW. This made the power system to disintegrate into several islands to make its operations healthy and stable. At the instant of the disintegration of the power system, it is very important to balance the reactive power consumption in the National Grid. This can be achieved by:

- Tripping transmission lines (e.g. KHD-Kelaniya1, KHD-Keleniya2 etc.), which do not contribute to cater power to any load connected.
- Generators themselves through reactive power compensation.
- The use of SVC (if available)

The generators which were in continuous operation in the National Grid are appeared in Table 4.9, with their generations immediately before and after the disintegration of the power system.

Table 4.10 shows the loads connected to each grid substation, which were involved in the Phase-I of the Load Shedding scheme. During the operation of the Phase-I of the LS scheme, 30% of the total load (or demand $\approx$ 498.58 MW) of the power system was considered.

Table 4.11 presents the 40% of the total demand that should be given priority in catering the load without any interruption. This is the load that is considered as the load in the Nation Grid, at the instant of the disintegration of the power system.

Further, the Stage, at which the 'disintegration of the power system' occurs, has been named as '*Stage 7*'.

From the instant of disintegration of the power system into several islands, each island starts its operation/catering the load, as an independent small power system.

Table 4.9 Generators in continuous operation in the National Grid immediately before and after the disintegration of the power system

Generator	Power generation (MW)	
	Immediately before disintegration of the PS	Immediately after disintegration of the PS
Sapugaskanda1	17.69	17.7
Sapugaskanda2	17.69	17.7
Sapugaskanda3	35.38	35.39
Upper-Kotmale	69.17	69.28
Kotmale	69.71	69.95
NewLaxapana1	19.61	19.61
NewLaxapana2	19.6	19.62
Laxapana1	24.92	24.96
Laxapana2	15.63	15.68
Polpitiya1	28.03	28.07
Polpitiya2	27.76	27.81
Victoria1	38.51	38.61
Victoria2	55.85	67.04
Victoria3	38.52	38.62
Wimalasurendra	23.48	23.5
Randenigala	44.5	44.57
GenPuttalam2	47.58	47.36
Total	593.63	605.47

To initiate its work, it is very important to balance the active and reactive power generation and demand within the island. First, active power balance between generation and demand is done by shedding the excess amount of demand, than the prevailing generation. This is implemented at the '*Stage 7*'. Simultaneously, the reactive power balance also should be done. A sample calculation has been done for island Matugama:

With the stage7,

Reactive power available (from generators) $= 46.84$ MVAR

Reactive power in demand $= 71.3 - (24.3 + 28.17) = 18.83$ MVAR

$Q_{Panadura}$ $Q_{Horana(a\ part)}$

∴ Excess reactive power generated $= 46.84 - 18.83 = 28.01$ MVAR

Generated–inductive Load–inductive

$= 28.01$ MVAR

Table 4.10 Loads connected to each grid substation, which were involved in the Phase-I of the load shedding scheme

Grid substation	Demand	
	Active power (MW)	Reactive power (MVAR)
Kiribathkumbura	51	21.1
Matara	30	13.6
Embilipitiya	14	4.3
Anuradhapura	14.3	4.6
Hambantota	14.2	6.7
Beliatta	8	1.9
Aturugiriya	24	16
Kolonnawa	60.6	41.1
Kosgama	39	25.1
Col-I	39	26.9
Col-A	49.3	34.5
Dehiwala	28.6	16.8
Ratmalana	49.5	41.3
J-Pura	16.5	7.8
Pannipitiya	52.58	35.3
Total	490.58	297

∴ To balance the reactive power Q, an additional capacitor with capacitive reactance (=28.01 MVAR) should be connected with the stage 7 (at the instant of disintegration of the power system) as shown in Table 4.12.

In the same manner reactive power compensation can be done for all the other islands to ensure a stable operation within each island. Practically this can be achieved through Static VAR Compensation (by installing SVCs). From Tables 4.13, 4.14 and 4.15 presents the Inductive/capacitive reactance required for reactive power compensation within the islands Rantembe, Embilipitiya and Kiribathkumbura respectively.

During disintegration of the power system, balance between demand and generation of active and reactive power within each island is equally important. In order to maintain balance between generation and demand, it is the practice to shed the required amount of load. By shedding these loads, certain transmission lines may be left energized but without catering any loads/load-centres. This can lead for instability situations (brown-outs or black-outs) of the power system due to capacitive reactance. So it is very important to isolate the transmission lines by tripping them off from both ends, if they do not cater any loads/load-centres.

The Phase-II of the proposed load shedding scheme is implemented sequentially. This operation is specific to each island. In general, the amount of load that involves in the Load Shedding Scheme-II is 10% (24.7 × 10% = 2.47 MW) of the available generation. Out of this 10%, the Load Shedding Scheme gets implemented with

Table 4.11 Loads connected to each substation in the National Grid, with a 40% of the total demand

Grid substation	Demand	
	Active power (MW)	Reactive power (MVAR)
Thulhiriya	37	14.5
Oruwala	0.0	0.0
Seethawaka	15	10.5
Pannipitiya	10.42	7
Kelaniya	21.1	15
Col-E	46	36
Col-F	38	17.4
Kelanitissa	17	11.5
Sub-C	24.2	14.5
Sapugaskanda	63.3	45.3
Biyagama	70	30
Aniyakanda	30	18
Kotugoda	71.2	25.4
Veyangoda	32.2	22.8
Katunayake	41.4	19.8
Bolawatta	51	35.6
Pannala	33.8	21.8
Madampe	36.2	21.7
Puttalam (Heladanavi)	22	3.7
Wimalasurendra	1.0	2.0
Total	660.82	372.5

Table 4.12 Inductive reactance required to be connected for reactive power compensation in Island Matugama

Island Matugama				
Grid	Generation		Demand	
	Active (MW)	Reactive (MVAR)	Active (MW)	Reactive (MVAR)
Ambalangoda	–	–	19.8	11.2
Matugama	–	–	33.9	7.4
Horana	–	–	2.64+	1.87+
Panadura	–	–		
Kukule	28.17 × 2 = 56.34	23.42 × 2 = 46.84	–	–
Total	56.34	46.84	56.34	20.47

∴ For reactive power compensation, inductive reactance required = 46.84 − 20.47 = 26.37 MVAR

25, 25, and 50% in stages 1, 2 and 3 respectively. The activation of the number of steps totally depends on the state of the island.

Table 4.13 Inductive reactance required to be connected for reactive power compensation in Island Rantembe

Island Rantembe

Grid	Generation		Demand	
	Active (MW)	Reactive (MVAR)	Active (MW)	Reactive (MVAR)
Rantembe	24.17	14.6	5.1	3.2
Badulla	–	–	18	7.6
Nuwara Eliya	–	–	1.07+	0.39+
Ampara	–	–		
Total	24.17	14.6	24.17	11.19

∴ For reactive power compensation, inductive reactance required = 14.6 – 11.19 = 3.41 MVAR

Table 4.14 Inductive reactance required to be connected for reactive power compensation in Island Embilipitiya

Island Embilipitiya

Grid	Generation		Demand	
	Active (MW)	Reactive (MVAR)	Active (MW)	Reactive (MVAR)
Balangoda	–	–	2.0	0.3
Ratnapura	–	–	0.2	6.1
Deniyaya	–	–	6.8	2.8
Galle	–	–	100.38+	9.99+
Embilipitiya	37.05	24.36		
Hambantota	–	–		
Beliatta	–	–		
Matara	–	–		
Samanalawewa	36.15 + 36.18 = 72.33	9 + 9 = 18	–	–
Total	109.38	42.36	109.38	19.19

∴ For reactive power compensation, inductive reactance required = 42.36 – 19.19 = 23.17 MVAR

A sample calculation which corresponds to Active Power of the 'Island Rantembe' is present below:

Total generation within the Island-Rantembe at the instant of disintegration of the power system	=24.17 MW
Power in demand at that instant	=66.3 MW
∴ the excess demand	=42.13 MW
Amount of load involved in the load shedding scheme	=24.17 MW × 10% = 2.417 MW
∴ amount of load to be shed in step 2	=2.417 MW × 25% = 0.604 MW
∴ amount of load to be shed in step 3	=2.417 MW × 25% = 0.604 MW
∴ amount of load to be shed in step 4	=2.417 MW × 50% = 1.208 MW

Table 4.15 Inductive reactance required to be connected for reactive power compensation in Island Kiribathkumbura

Island Kiribathkumbura				
Grid	Generation		Demand	
	Active (MW)	Reactive (MVAR)	Active (MW)	Reactive (MVAR)
Kiribathkumbura	–	–		
Ukuwela	20.25	12.42		
Bowatenna	8.43	5.36	–	–
Habarana	–	–		
Valachchena	–	–	11.7	6.5
Kurunegala	–	–		
Anuradhapura	–	–		
N. Anuradhapura	–	–		
Trincomalee	–	–		
Vavunia	–	–	11.6	3.3
Kilinochchi	–	–	+5.38	+0.56
Total	28.68	17.78	28.68	10.36

∴ For reactive power compensation, inductive reactance required = 17.78 − 10.36 = 7.42 MVAR

Table 4.16 and Fig. 4.9 demonstrate the sequential steps that correspond to 'Island Rantembe' which operates as an island in the Phase II of the proposed load shedding scheme.

Table 4.17 provides the excess demand and amounts of loads to be shed in shedding stages corresponding to each island at the disintegration of the power system.

Table 4.16 Phase-II of proposed load shedding scheme-II with time delays corresponding to Island Rantembe

Frequency state	Load to be shed		Time delay
f ≤ 48.6 Hz	Load equivalent to the difference between generation and demand (42.13 MW)	66.3 − 24.17 = 42.13 MW	Step 1: 200 ms
f < 48.4 Hz	Shed a load of 25% from a 10% of the generation at the instant of islanding	(24.17 × 10%) × 25% = 0.604 MW	Step 2: 200 ms
f < 48.2 Hz	Shed a load of 25% from a 10% of the generation at the instant of islanding	(24.17 × 10%) × 25% = 0.604 MW	Step 3: 200 ms
f < 48.0 Hz	Shed a load of 50% from a 10% of the generation at the instant of islanding	(24.17 × 10%) × 50% = 1.208 MW	Step 4: 200 ms

Fig. 4.9 Flow chart of the Phase-II of load shedding scheme-II that corresponds to Island-Rantembe. The implementation of the number of Load Shedding stages is sequential and specific to this Island-Rantembe

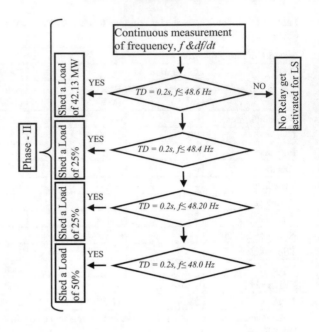

Table 4.17 The excess demand and amounts of loads to be shed in shedding stages in each island during disintegration of the power system

Island	Frequency state (Hz)	Load to be shed (MW)	Time delay (ms)
Island Rantembe	f ≤ 48.6	42.13 (excess demand)	Step 1: 200
	f < 48.4	0.604	Step 2: 200
	f < 48.2	0.604	Step 3: 200
	f < 48.0	1.208	Step 4: 200
Island Matugama	f ≤ 48.6	44.31 (excess demand)	Step 1: 200
	f < 48.4	1.36	Step 2: 200
	f < 48.2	1.36	Step 3: 200
	f < 48.0	2.71	Step 4: 200
Island Embilipitiya	f ≤ 48.6	14.4 (excess demand)	Step 1: 200
	f < 48.4	2.74	Step 2: 200
	f < 48.2	2.74	Step 3: 200
	f < 48.0	5.47	Step 4: 200
Island Kiribathkumbura	f ≤ 48.6	134 (excess demand)	Step 1: 200
	f < 48.4	0.72	Step 2: 200
	f < 48.2	0.72	Step 3: 200
	f < 48.0	1.44	Step 4: 200

After the controlled disintegration of the national grid, the frequency and voltage responses of all four islands and the national grid were observed. They all were within specified limits.

Chapter 5
Results and Discussion

The Load Shedding Schemes described in Table 4.2 (CEB Load Shedding Scheme) and Tables 4.4, 4.5 and 4.6 (Proposed Load Shedding Schemes) were simulated in the Power System of Sri Lanka. Different scenarios of generation deficits were considered to observe the frequency and voltage responses of the schemes.

5.1 Discussion: Load Shedding Scheme-I with Generation Deficit of 829.6 MW

The highest generation capacity of a single generator in Sri Lanka is 300 MW. Out of this 25 MW is consumed by the power plant itself for its auxiliary services. So it can be considered the capacity of such generator is 275 MW. Since Sri Lanka processes 03 nos. such generators, the behavior of the power system during a generation deficit of 829.6 MW (which is equivalent to an outage of all three coal power plants = 3×275 MW) was implemented in the simulation model. Figure 5.1 shows the frequency profile of the power system with the implementation of the Load Shedding Scheme-I. The frequency reaches a steady state condition of 50.39 Hz after the disturbance.

Figure 5.2 demonstrates the voltage profile of the power system after the disturbance. The steady state voltage is 217.6 kV, which falls within safe limits of the voltage range of the power system, $220 \pm 5\%$ kV.

The rate of change of frequency variation after the disturbance is shown in Fig. 5.3. The minimum frequency degradation rate is -1.51 Hz/s.

Figure 5.4 shows some of the stages involved in the Load Shedding Scheme. In order to address this disturbance, all six stages of the LSS got implemented.

© Springer Nature Singapore Pte Ltd. 2018   121
T. Bambaravanage et al., *Modeling, Simulation, and Control of a Medium-Scale Power System*, Power Systems, https://doi.org/10.1007/978-981-10-4910-1_5

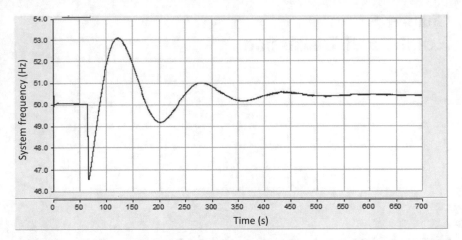

Fig. 5.1 Frequency profile with the implementation of the LSS-I in response to a generation deficit of 829.6 MW. Steady state frequency = 50.39 Hz; maximum frequency achieved = 53.06 Hz; minimum frequency achieved = 46.55 Hz

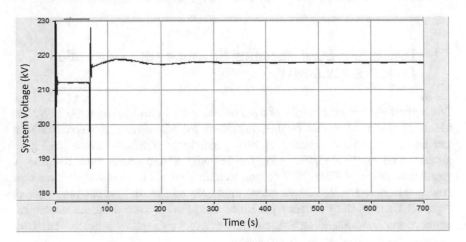

Fig. 5.2 Voltage profile of the national grid. Steady state voltage = 217.6 kV; maximum voltage achieved = 227.95 kV; minimum voltage achieved = 187.26 kV

Some of the generator performance at the instant of the disturbance is shown in Fig. 5.5. Hence with the implementation of the Proposed Load Shedding Scheme-I, the system can be brought to a steady state condition even during a large disturbance such as 825 MW (which is equivalent to tripping of the coal power plant), that can't be achieved by implementing the current load shedding scheme of the CEB.

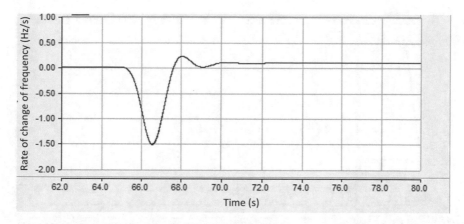

Fig. 5.3 Rate of change of frequency after the disturbance. Minimum value achieved = −1.51 Hz/s

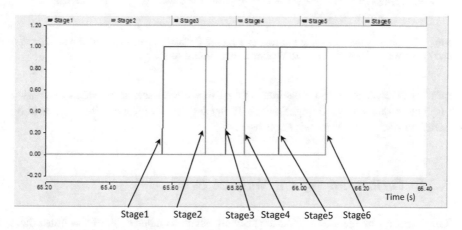

Fig. 5.4 Stages involved in the load shedding scheme

5.2 Discussion: Load Shedding Scheme-II with Generation Deficit of 495.14 MW

With reference to the frequency profile in Fig. 5.1, even though the system frequency get stabilized within stability limits (49.5 Hz ≤ f ≤ 50.5 Hz), with the implementation of the Load Shedding Scheme-I, the frequency goes beyond 53 Hz for few seconds (4 s) and reaches below 47 Hz for ≈ 2.5 s. Even though the maximum voltage the system reaches is within safe limits, the minimum voltage achieved in the system goes below 198 kV (for emergency situations allowable voltage range is 198 kV ≤ v ≤ 242 kV). The proposed Load Shedding scheme-II eliminates/minimizes such undesirable situations letting a chance for a further

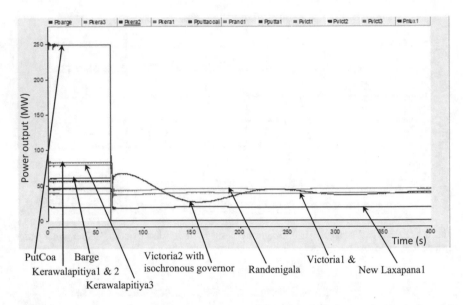

number of consumers to be catered. Out of the several scenarios which were tried out with the Load shedding Scheme-II, system performance with a generation deficit of 495.14 MW is discussed here.

5.2.1 Performance of the National Grid

The intention of the Load Shedding Scheme-II is to cater 40% of the consumers without any disturbance during a faulty situation in the power system (national grid). With a forced outage of 495.14 MW generation, due to the implementation of LSS-II, the national grid could be brought to a stable condition. Figure 5.6 shows the frequency profile of the national grid after the considering disturbance. The frequency always retains within limits: $48.0 \leq f \leq 52.0$ Hz. It reaches a steady state frequency of 49.75 Hz after the disturbance.

The voltage profile of the national grid is appeared in Fig. 5.7. For transmission voltage of 220 kV, it has a tolerance of 220 kV $\pm$ 5% during a disturbance, i.e. 209 kV $\leq v \leq 231$ kV and it can have a tolerance of 220 kV $\pm$ 10% during an emergency, i.e. 198 kV $\leq v \leq 242$ kV. The simulation results show that the voltage reaches a steady state condition within stable limits which equals 212.81 kV.

The rate of change of frequency is shown in Fig. 5.8. The minimum rate of change of frequency it reaches because of the disturbance is 0.793 Hz/s.

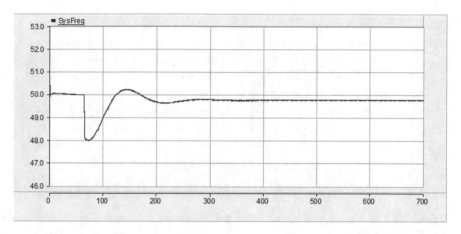

Fig. 5.6 Frequency profile of the national grid with the disturbance. Steady state frequency = 49.75 Hz

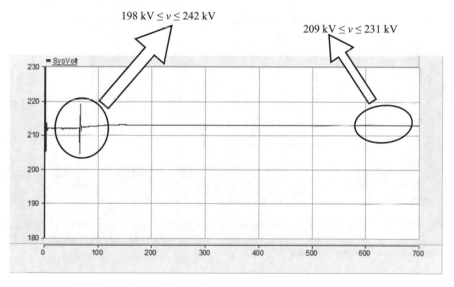

Fig. 5.7 Voltage profile of the national grid during and after the disturbance. Steady state voltage = 212.81 kV

Because of the forced outage of generation with a capacity of 495.14 MW, all four stages of the Phase-I of the LSS got implemented and it initiated the islanding operation of the Load Shedding Scheme, which is given by the '*Stage 7*'. Figure 5.9 shows the stages that got implemented and when they got implemented.

Figure 5.10 demonstrates the generation output of some generators that got tripped and some other generators which are catering the demand.

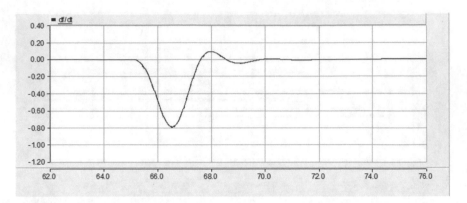

Fig. 5.8 Rate of change of frequency after the disturbance. Minimum rate of change of freq. = −0.793 Hz/s

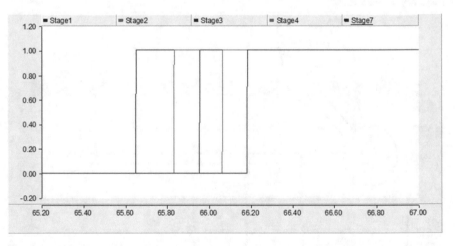

Fig. 5.9 Stages got implemented in the load shedding scheme. 4 nos. of stages of the phase-I and the disintegration of the power system. The stage 7 corresponds to the 5th stage which leads for disintegration of the power system

When consider the stability of the islands formed during the disintegration of the power system, it can be shown that the frequency and the voltage of each island retain within stability limits.

5.2.2   Performance of Island Rantembe

It is possible to maintain the stability of the Island Rantembe while catering part of the consumers in the island. Its control station is located in the grid Rantembe as

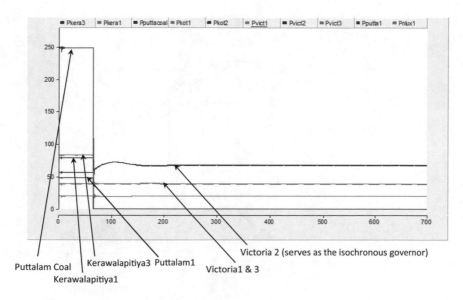

Fig. 5.10 Some of the power plants that got tripped-off and generation output of some selected generators

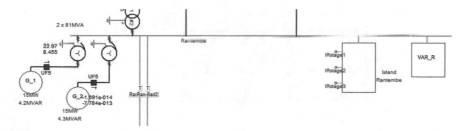

Fig. 5.11 Island Rantembe control station is located in the Grid-Rantembe

shown in Fig. 5.11. Until the disintegration of the power system occurs, the Generator Rantembe operates as one of the hydro-generators of the power system of Sri Lanka. But when disintegration is done, immediately it takes over the role of the isochronous governor. Figure 5.12 shows the control logic used to perform as a swing generator.

According to Fig. 5.13, the frequency profile of island Rantembe which occurs with the disintegration of the power system retains within specified limits: $52 \text{ Hz} \geq f \geq 48 \text{ Hz}$. The maximum and minimum frequencies it reached immediately after islanding is 51.45 and 48.24 Hz while the steady state frequency it reached is 50.08 Hz.

Figure 5.14 clearly shows the voltage during the islanding agrees the condition: $145.2 \text{ kV} \geq v \geq 118.8 \text{ kV}$. The maximum voltage it reached, 135 kV is less than the maximum system voltage allowable and the minimum voltage it reached,

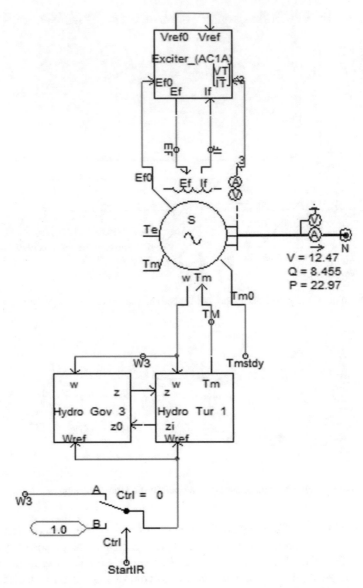

Fig. 5.12 Generator Rantembe takes over the role of the isochronous governor just after islanding operation occurs

124 kV is greater than the minimum system voltage allowable. The steady state voltage it reached is 130.38 kV which satisfies the condition $145.2 \geq v \geq 118.8$ kV.

The rate of change of frequency during the disintegration of the power system is shown in Fig. 5.15. It comes to a steady state condition with time.

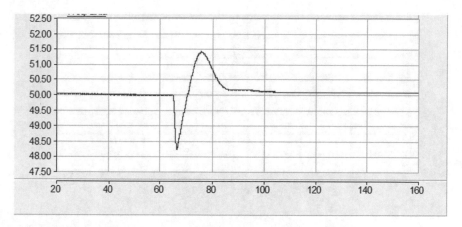

Fig. 5.13 Frequency profile of Island Rantembe

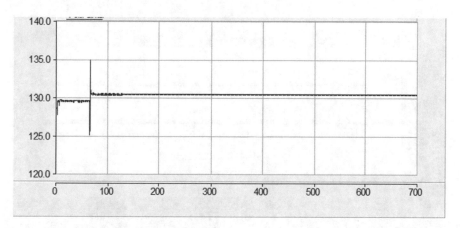

Fig. 5.14 Voltage profile of Island Rantembe

A load shedding scheme should be introduced to make the system balance during the transient period. In the Island Rantembe, during the disintegration of the power system 'stage 1' of the load shedding scheme which is known as 'IRstage1' got implemented. This is demonstrated in Fig. 5.16.

In order to balance the reactive power of the island an inductive load has been connected to the island (an action equivalent to the static var compensator—SVC) through a sequencer. The graph shows the instant the reactance got connected to the 'Island Rantembe—power system' which is the instant that the disintegration of the National Grid occurs. This is shown in Fig. 5.17. The control logic—'sequencer' used (in PSCAD) to connect the reactance to the power system (for reactive power compensation) and the inductor used for reactive power compensation are shown in Figs. 5.18 and 5.19 respectively.

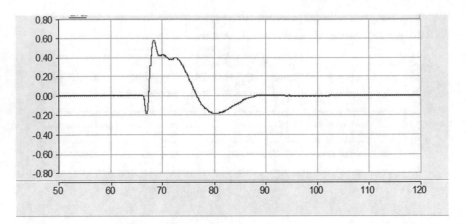

Fig. 5.15 Rate of change of frequency of Island Rantembe during disintegration of the power system

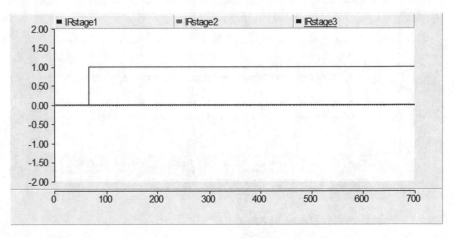

Fig. 5.16 'Stage 1' of the load shedding scheme of the Island Rantembe, known as 'IRstage1' got implemented to stabilize the system

Since the generator Rantembe has been connected to the swing bus, its active and reactive power generations after the islanding operation, are shown in Figs. 5.20 and 5.21 respectively.

5.2.3 Performance of Island Matugama

With the disintegration of the power system the simulation results show that the Island Matugama could be brought to a balance and stable power system. Its control

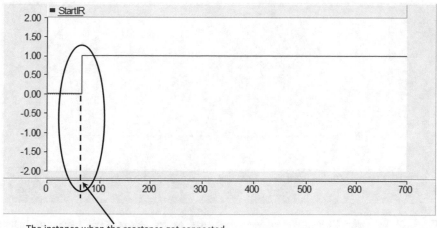

The instance when the reactance got connected

Fig. 5.17 At the instant the reactance got connected to the 'Island Rantembe—power system', the state of the graph goes high

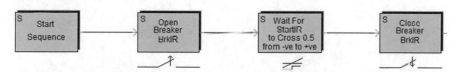

Fig. 5.18 The control logic—'sequencer' used (in PSCAD) to connect the reactance to the 'Island Rantembe—power system'

Fig. 5.19 Inductor used for reactive power compensation

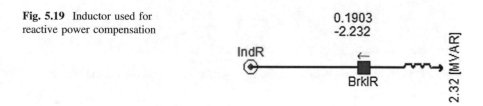

station is located in the grid Kukule as shown in Fig. 5.22. Until the disintegration of the power system occurs, the Generator Kukule1 operates as one of the hydro-generators of the power system of Sri Lanka. But when disintegration is done, immediately it takes over the role of the isochronous governor. Figure 5.23 shows the control logic used to perform as a swing generator.

The frequency profile of island Matugama with the disintegration of the power system, should retain within specified limits: 52 Hz $\geq$ f $\geq$ 48 Hz. But according to Fig. 5.24, the maximum and minimum frequencies it reached immediately after islanding is 52.2 and 48.24 Hz. Even though the maximum frequency exceeds the maximum specified frequency limit, i.e. 52.0 Hz, it lasts just for 10 s duration. The steady state frequency it reached is 50.19 Hz.

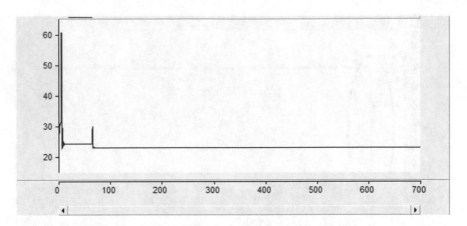

Fig. 5.20 Generator Rantembe has been considered as the generator with the isochronous governor. Its power generation against time, after islanding operation

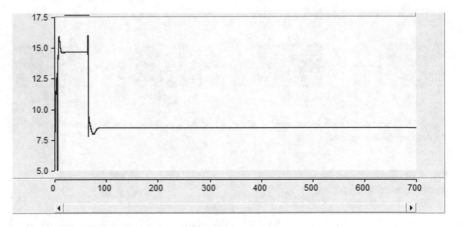

Fig. 5.21 Reactive power generation of the generator Rantembe with the isochronous governor

Figure 5.25 clearly shows the voltage during the islanding is 145.2 kV $\geq v \geq$ 118.8 kV. The maximum voltage it reached, 132.97 kV is less than the maximum system voltage allowable and the minimum voltage it reached, 123.04 kV is greater than the minimum system voltage allowable. The steady state voltage it reached is 129.15 kV which satisfies the condition 145.2 $\geq v \geq$ 118.8 kV.

The rate of change of frequency during the disintegration of the power system is shown in Fig. 5.26. It comes to a steady state condition with time.

A load shedding scheme should be introduced to make the system balance during the transient period. In the Island Matugama, during the disintegration of the power system 'stage 1' of the load shedding scheme which is known as 'IMstage1' got implemented. This is demonstrated in Fig. 5.27.

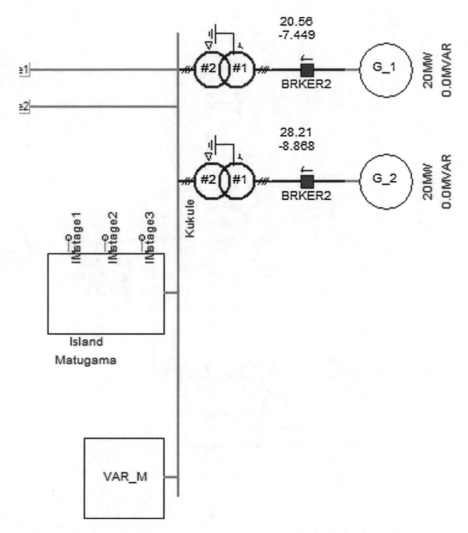

Fig. 5.22 Island Matugama control station is located in the Grid-Kukule

In order to balance the reactive power of the island an inductive load has been connected to the island (an action equivalent to the static var compensator—SVC) through a sequencer. The graph shows the instant the reactance got connected to the 'Island Matugama—power system' which is the instant that the disintegration of the National Grid occurs. This is shown in Fig. 5.28. The control logic—'sequencer' used (in PSCAD) to connect the reactance to the power system (for reactive power compensation) and the inductor used for reactive power compensation are shown in Figs. 5.29 and 5.30 respectively.

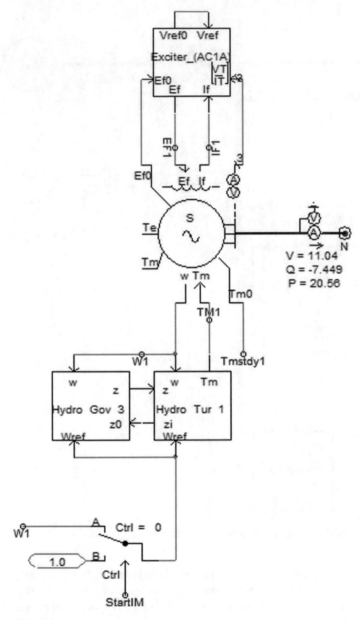

Fig. 5.23 Generator Kukule1 takes over the role of the isochronous governor just after islanding operation occurs

Since the generator Kukule1 has been connected to the swing bus, its active and reactive power generations after the islanding operation, are shown in Figs. 5.31 and 5.32 respectively.

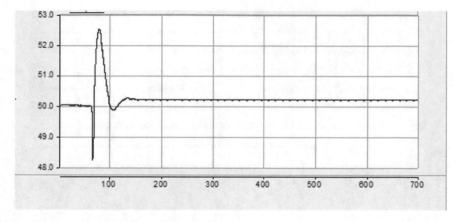

Fig. 5.24 Frequency profile of Island Matugama

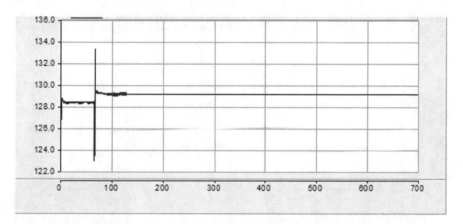

Fig. 5.25 Voltage profile of the Island Matugama

5.2.4   Performance of Island Embilipitiya

While catering part of the consumers in the island, the stability of the Island Embilipitiya can be maintained within specified frequency and voltage limits. Its control station is located in the grid-Samanalawewa as shown in Fig. 5.33. Until the disintegration of the power system occurs, the Generator Samanalawewa1 operates as one of the hydro-generators of the power system of Sri Lanka. But when disintegration is done, immediately it takes over the role of the isochronous governor. Figure 5.34 shows the control logic used for the Generator Samanalawewa1 to perform as a swing generator.

The frequency profile of island Embilipitiya with the disintegration of the power system, should retain within specified limits: $52 \text{ Hz} \geq f \geq 48 \text{ Hz}$. But according

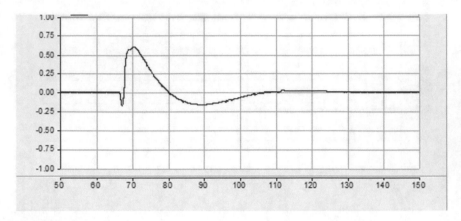

Fig. 5.26 Rate of change of frequency after the disturbance in Island Matugama

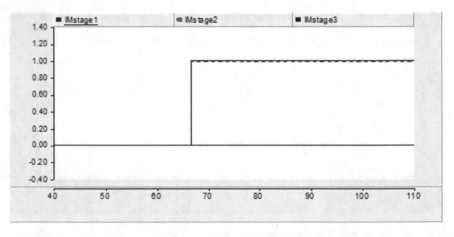

Fig. 5.27 'Stage 1' of the load shedding scheme of the Island Matugama, known as 'IMstage1' got implemented to make the system stable

to Fig. 5.35, the maximum and minimum frequencies it reached immediately after islanding is 53.76 and 48.39 Hz. But during the islanding operation, the frequency exceeds the specified limits and it lasts for 25 s beyond 52.0 Hz. The steady state frequency it reached is 50.4 Hz.

Figure 5.36 clearly shows the voltage during the islanding is 145.2 kV $\geq v \geq$ 118.8 kV. The maximum voltage it reached, 133.2 kV is less than the maximum system voltage allowable and the minimum voltage it reached, 120.95 kV is greater than the minimum system voltage allowable. The steady state voltage it reached is 130.02 kV which satisfies the condition 145.2 $\geq v \geq$ 118.8 kV.

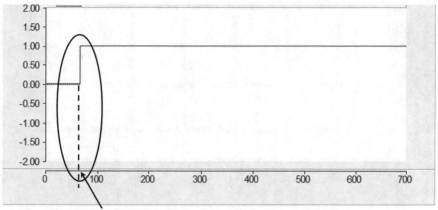

The instance when the reactance got connected

Fig. 5.28 At the instant the reactance got connected to the 'Island Matugama—power system', the state of the graph goes high

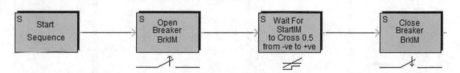

Fig. 5.29 The control logic—'sequencer' used (in PSCAD) to connect the reactance to the 'Island Matugama—power system'

Fig. 5.30 Inductor used for reactive power compensation

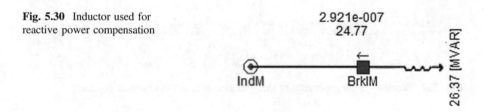

The rate of change of frequency during the disintegration of the power system is shown in Fig. 5.37. It comes to a steady state condition with time.

The load shedding scheme which was introduced to make the system balance during the transient period which occur after the disintegration of the power system, got activated by implementing the 'stage 1' which is known as 'IEstage1'. This is demonstrated in Fig. 5.38.

In order to balance the reactive power of the island an inductive load has been connected to the island (an action equivalent to the static var compensator—SVC) through a sequencer. The graph shows the instant the reactance got connected to the

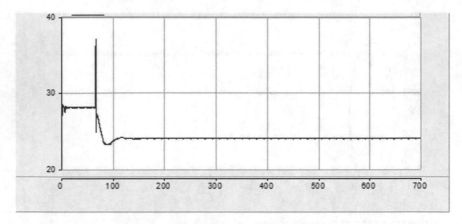

Fig. 5.31 Generator Kukule1 has been considered as the generator with the isochronous governor. Its power generation against time after islanding operation

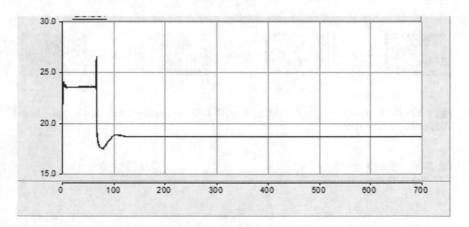

Fig. 5.32 Reactive power generation of the generator with the isochronous governor

'Island Embilipitiya—power system' which is the instant that the disintegration of the National Grid occurs. This is shown in Fig. 5.39. The control logic—'sequencer' used (in PSCAD) to connect the reactance to the power system (for reactive power compensation) and the inductor used for reactive power compensation are shown in Figs. 5.40 and 5.41 respectively.

Since the generator Samanalawewa1 has been connected to the swing bus, its active and reactive power generations after the islanding operation, are shown in Figs. 5.42 and 5.43 respectively.

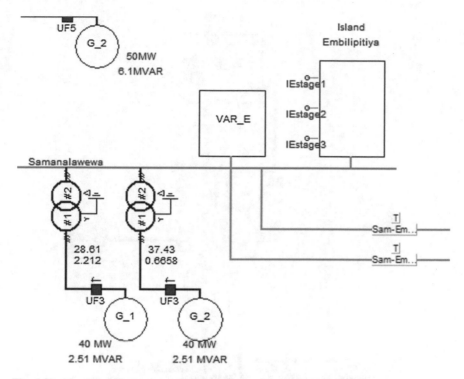

Fig. 5.33 Island Embilipitiya control station is located in the Grid-Samanalawewa

5.2.5 Performance of Island Kiribathkumbura

While catering part of the consumers in the island, the stability of the Island Kiribathkumbura can be maintained within specified frequency and voltage limits. Its control station is located in the grid Ukuwela as shown in Fig. 5.44. Until the disintegration of the power system occurs, the Generator Ukuwela operates as one of the hydro-generators of the power system of Sri Lanka. But when disintegration is done, immediately it takes over the role of the isochronous governor. Figure 5.45 shows the control logic used for the Generator Ukuwela to perform as a swing generator.

According to Fig. 5.46, the frequency profile of island Kiribathkumbura which occurs with the disintegration of the power system retains within specified limits: 52 Hz $\geq f \geq$ 48 Hz. The maximum and minimum frequencies it reached immediately after islanding is 50.24 and 47.65 Hz while the steady state frequency it reached is 49.88 Hz.

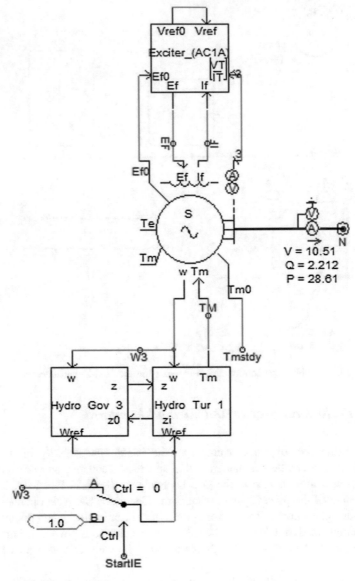

Fig. 5.34 Generator Samanalawewa1 takes over the role of the isochronous governor just after islanding operation occurs

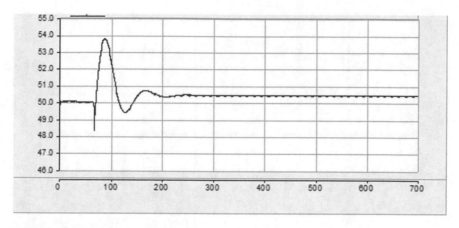

Fig. 5.35 Frequency profile of Island Embilipitiya

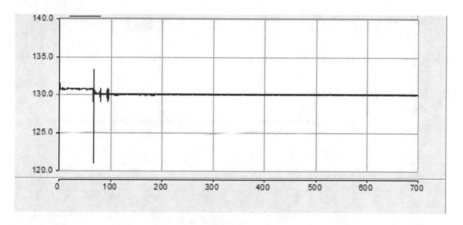

Fig. 5.36 Voltage profile of Island Embilipitiya

Figure 5.47 clearly shows the voltage during the islanding agrees the condition: 145.2 kV $\geq v \geq$ 118.8 kV. The maximum voltage it reached, 142.78 kV is less than the maximum system voltage allowable and the minimum voltage it reached, 125.75 kV is greater than the minimum system voltage allowable. The steady state voltage it reached is 132.31 kV which satisfies the condition $132 \pm 10\%$ kV ($145.2 \geq v \geq$ 118.8 kV).

The rate of change of frequency during the disintegration of the power system is shown in Fig. 5.48. It comes to a steady state condition with time.

The load shedding scheme which was introduced to make the system balance during the transient period which occurs after the disintegration of the power system, got activated by implementing the 'stage 1' and 'stage 2' which are known as 'IKstage1' and 'IKstage2'. This is demonstrated in Fig. 5.49.

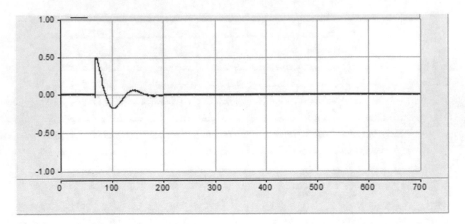

Fig. 5.37 Rate of change of frequency after the disturbance in Island Embilipitiya

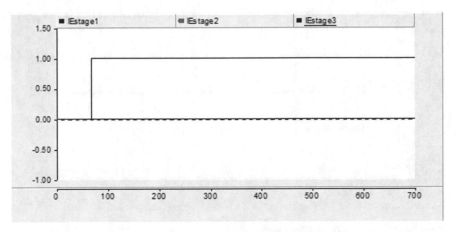

Fig. 5.38 'Stage 1' of the load shedding scheme of the Island Embilipitiya, known as 'IEstage1' got implemented to make the system stable

In order to balance the reactive power of the island an inductive load has been connected to the island (an action equivalent to the static var compensator—SVC) through a sequencer. The graph shows the instant the reactance got connected to the 'Island Kiribathkumbura—power system' which is the instant that the disintegration of the National Grid occurs. This is shown in Fig. 5.50. The control logic —'sequencer' used (in PSCAD) to connect the reactance to the power system (for reactive power compensation) and the inductor used for reactive power compensation are shown in Figs. 5.51 and 5.52 respectively.

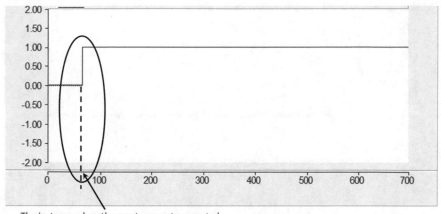

The instance when the reactance got connected

Fig. 5.39 At the instant the reactance got connected to the 'Island Embilipitiya—power system', the state of the graph goes high

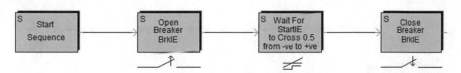

Fig. 5.40 The control logic—'sequencer' used (in PSCAD) to connect the reactance to the 'Island Embilipitiya—power system'

Fig. 5.41 Inductor used for reactive power compensation

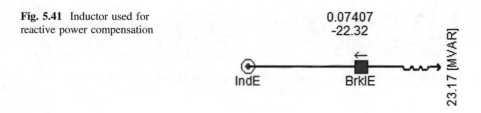

Since the generator Ukuwela has been connected to the swing bus, its active and reactive power generations after the islanding operation, are shown in Figs. 5.53 and 5.54 respectively.

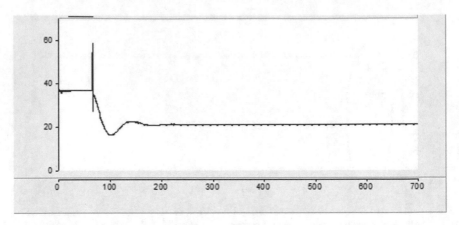

Fig. 5.42 Generator Samanalawewa1 has been considered as the generator with the isochronous governor. Its power generation against time after islanding operation

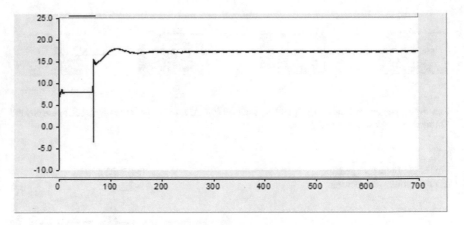

Fig. 5.43 Reactive power generation of the generator with the isochronous governor

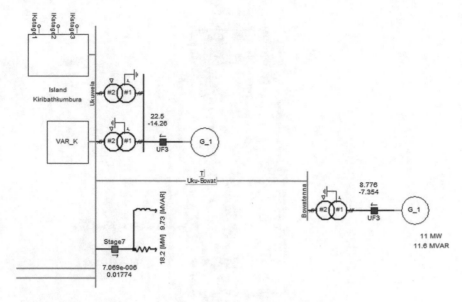

Fig. 5.44 Island Kiribathkumbura control station is located in the Grid-Ukuwela

5.3   Performance Comparison on Selected Load Shedding Schemes (LSS)

A comparison of performance of the three load shedding schemes:

- CEB Load Shedding Scheme
- Proposed Load Shedding Scheme-I
- Proposed Load Shedding Scheme-II, was carried out simulating different scenarios on the power system of Sri Lanka (Simulation model done with PSCAD EMTDC).

Each scenario was applied to all three schemes and several such scenarios were considered. Generation deficits were originated by forced capacity outages and corresponding frequency variations occurred for the LSSs were tabulated as shown in Table 5.1 for comparison. The observations were received from a Power System simulation done for a load flow with actual data, received from the CEB. A model of the Power System of Sri Lanka was simulated using the software PSCAD/EMTDC, at that instant the total generation was 1637 MW. The results discussed here are based on the above total generation.

With the results obtained in this simulation, it is evident that a LS step which gets implemented after 48.6 Hz can lead to situations where SF goes below 47.5 Hz. This can also lead to instability situations in the PS.

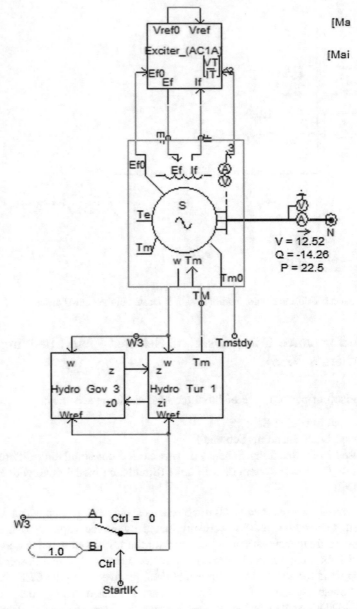

Fig. 5.45 Generator Ukuwela takes over the role of the isochronous governor just after islanding operation occurs

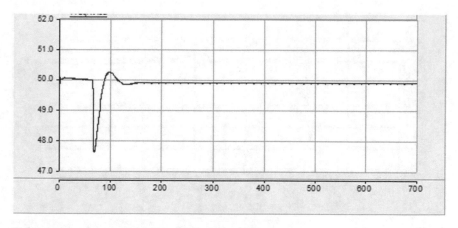

Fig. 5.46 Frequency profile of Island Kiribathkumbura

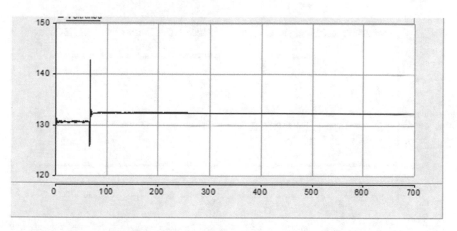

Fig. 5.47 Voltage profile of Island Kiribathkumbura

- In the CEB LSS, the adaptive LS step (amounts to 21% of the total load) is to get implemented at df/dt ≤ 0.85 Hz/s and SF = 49.0 Hz. As the maximum ROCOF occurs after about 1.5 s from the instant of the initiation of the disturbance, for such a large disturbance this 0.85 Hz/s is achieved after 48.6 Hz. Before reaching this state, possibility of implementing earlier steps can lead to excessive LS. Hence the effectiveness of the LS step is questionable.

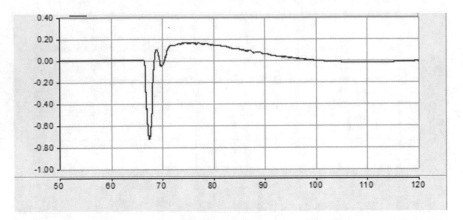

Fig. 5.48 Rate of change of frequency of Island Kiribathkumbura during disintegration of the power system

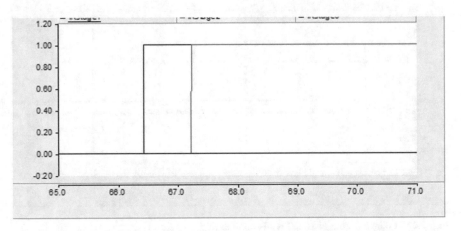

Fig. 5.49 'Stage 1' and 'stage 2' of the load shedding scheme of the Island Kiribathkumbura, known as 'IKstage1' and 'IKstage2' respectively got implemented to make the system stable

- The LSS itself has a step (5th step) with the condition for implementing if frequency <47.5 Hz with no intentional time delay. Even if that step gets implemented, the PS takes further time to bring the SF above 47.5 Hz. In Sri Lanka, thermal PPs contribute to a considerable portion of power generation, which is ⪅65% of its average total generation. Hence it has a large tendency to collapse the whole PS with the available LSS.

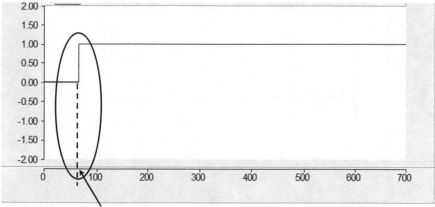

The instance when the reactance got connected

Fig. 5.50 At the instant the reactance got connected to the 'Island Kiribathkumbura—power system', the state of the graph goes high

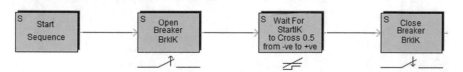

Fig. 5.51 The control logic—'sequencer' used (in PSCAD) to connect the reactance to the 'Island Kiribathkumbura—power system'

Fig. 5.52 Inductor used for reactive power compensation

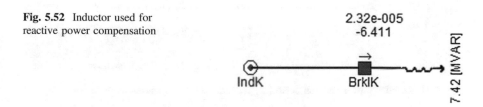

- The CEB LSS helps to maintain the stability of the PS only at certain amounts of generation deficits. The frequency profile with 275 MW capacity outage, corresponding to tripping of Lakvijaya Coal PP (one generator—since each generator contributes the PS with 275 MW constant power), clearly shows that the system becomes stable with the current LSS. But if 2 nos. or 3 nos. coal PPs get tripped that may lead even to a catastrophic failure.
- Therefore it is very important to go for a LSS which does not lead the system frequency below the minimum allowable operating frequency (47.5 Hz) as well as that does not shed excessive load.

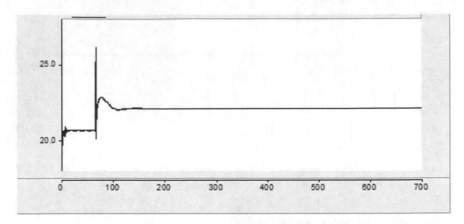

Fig. 5.53 Generator Ukuwela has been considered as the generator with the isochronous governor. Its power generation against time after islanding operation

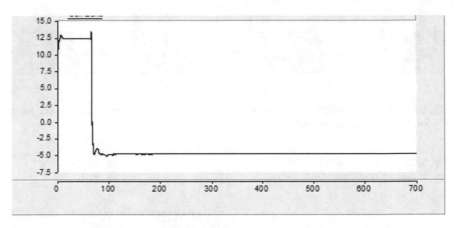

Fig. 5.54 Reactive power generation of the generator with the isochronous governor

- The PLSSs supports the PS during generation deficits due to tripping of small generators as well as large capacity PPs (e.g. Lakvijaya Coal PP-1, PP-2, PP-3 or any combination of these).
- The Proposed LSS-I can be implemented with the prevailing facilities in the PS of Sri Lanka. This suggests a better solution for the stability problem due to generation deficit.

Table 5.1 Simulation results of present CEB LSS, Proposed LSS-I and Proposed LSS-II when applied for the Power System of Sri Lanka, under different forced power generation outages

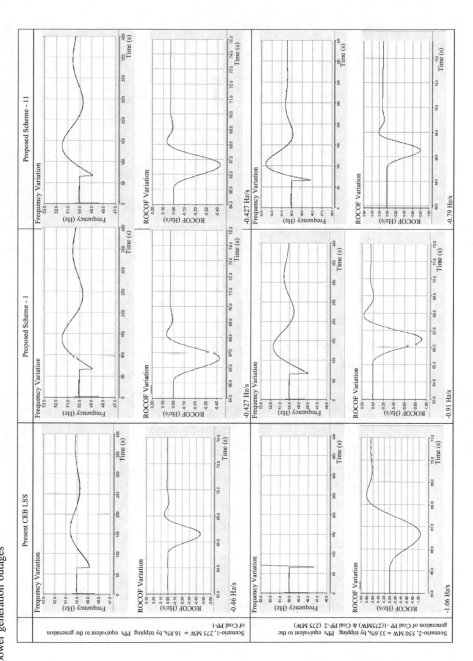

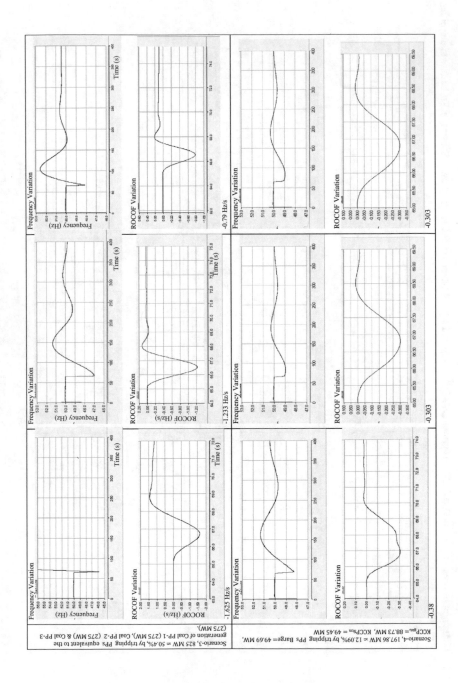

- If the generation capacity of Sri Lanka increases further with large generators and with fluctuating generation capacities, it is a good option to go for disintegration of the power system. This let more consumers have electricity with lesser interruptions. This is demonstrated in the Proposed LSS-II. Even though it suggests 40% of the load to give priority in catering electricity (when concern the national grid) further amount of load receives power due to islanding operation.

Chapter 6
Conclusion

This research was carried out based on a model of the power system of Sri Lanka, which was simulated suing PSCAD EMTDC and COMFORTRAN software. This power system simulation was done with reference to the load flow occurred on the 13th May, 2013 as the day-time peak demand, (around 12.30 p.m.). Different actual power system scenarios were simulated in this model and its results/performance was very much closer to real time values. But as I see, following are some points that should be accounted for, through which the performance of the simulation model could have been improved further.

- The power system is comprised with a variety of generator/turbine types, such as steam-turbine power plants, gas-turbine power plants and diesel engines. In simulating these thermal generators in the power system, I have considered all of them (thermal generators) as steam turbine power plants (Sect. 3.2.6).
- Even though the generation capacities are different I have used the same set of parameters in certain situations (especially in simulating the governor, turbine and generator) for thermal plants and hydro plants (two separate sets for the two corresponding types) (Sects. 3.2.4, 3.2.6 and 3.2.7, Tables 3.13 and 3.16).
- It was not possible to get the actual inertia values for all generators, for the simulation. I used some experimental values as well as some calculated values in this process, which may not be the exact vales of the inertias of the units connected to the national grid at that instant (Sect. 3.2.4, Table 3.10).
- In the simulation, the loads connected to bus bars at each grid were considered as fixed values. Even though the inductive and capacitive reactance varies with the system frequency variations, the option of inductive and capacitive loads which are not sensitive to system frequency was selected in the simulation.
- If we consider the construction details of a particular standard type underground cable, their construction details may be slightly different from one manufacturer to another. So the same values may not be applied in the simulation process, which are identical to the power system components.

© Springer Nature Singapore Pte Ltd. 2018
T. Bambaravanage et al., *Modeling, Simulation, and Control of a Medium-Scale Power System*, Power Systems, https://doi.org/10.1007/978-981-10-4910-1_6

- Devices such as CFL lamps, Variable Speed Drives, Switching devices etc. are becoming much popular in the country. Even though they have many advantageous situations to the consumer, on the other hand they introduce lot of harmonics whose effects may be very bad for the utility as well as the power system. This issue has not been addressed in the simulation process.

With the developed simulation model of the Power System of Sri Lanka, it was able to come to certain conclusions which are much favorable to maintain the system with a high quality and reliable service. The results show that the Proposed Load Shedding Schemes are better solutions for the power system stability problem during generation deficiencies. These proposed load shedding schemes are exclusively specific for the power system of Sri Lanka. It depends on the electrical power system practice, regulations, largest generator capacity/capacities, electricity consumption pattern, capacity of embedded generation etc.

- By implementing a load shedding scheme at an initial stage of a generation deficiency, further reduction of system frequency can be eliminated. This would be very supportive in regulating frequency in a power system. Accordingly the proposing frequency for initiating a Load Shedding Scheme is 49.4 Hz. With reference to the current practice of Sri Lanka, CEB initiates its Load Shedding Scheme if the system frequency < 48.75 Hz with a delay time of 100 ms.
- The chance of occurring the conditions "f < 49.0 Hz" and "−0.85 < df/dt" (which can be considered as adaptive to the system behavior) together (i.e. the logic condition 'f < 49.0 Hz and −0.85 < df/dt'), that are given in the CEB load shedding scheme is very much less. Hence there is a high possibility of initiating 'other stages' of the load shedding scheme in addition to this stage, which may lead to 'over-shedding'.
- If it is possible to limit implementing the load shedding stages in the Load Shedding Scheme only up to a system frequency of 48.6 Hz, the power system can be retained within the specified frequency limits (minimum safe operating frequency limit of thermal generator): i.e. the system frequency $\geq$ 47.0 Hz. Since Sri Lanka is a country which receives electricity mostly from thermal power generation, it is very important to keep the system frequency beyond the safe limit 47.0 Hz. Else this can lead for a catastrophic failure.
- Further by implementing disintegration of the national grid at an instant where the system frequency = 48.6 Hz, rather than considering a specific df/dt value, it is possible to maintain the frequencies in the national grid as well as in islands approximately above 47.5 Hz. There by the stability of the grid network can be assured.
- During disintegration of the power system it is very important to identify and isolate transmission lines which do not cater power to any loads but still energized while being connected to the power system. It is because this type of transmission lines can affect the reactive power balance of the power system and hence may lead to catastrophic failures.

With the help of the simulation program I was able to explain real time situations/problems which were experienced by engineers in the CEB,

- while they were trying to isolate the southern part of the national grid
- during a generation tripping occurred in the Laxapana and New-Laxapana power stations.

This was because reactive power imbalances occur in the National Grid due to No-Load or Lightly Loaded transmission lines in the transmission network.

- With the Proposed LSS-II, even though it suggests 40% of the load to give priority in catering electricity (when concern the national grid) further amount of load receives power due to islanding operation. Therefore, it is possible to cater a larger number of consumers by disintegration of the power system.
- In this simulation model SVC were not used; instead by doing a small calculation for reactive power consumption in each island, the reactive power compensation was done. More successful results can be assured if SVCs were introduced to each and every island as well as to the national grid.

Appendices
Appendix A
Transmission Lines

Power system's main power corridor: Transmission lines and transformers with a two-port network.

Referring to Fig. A.1a approximated two port transmission network and (b) phasor diagram:

V   receiving end phase voltage
E   sending end phase voltage
P   single-phase real power
Q   single-phase reactive power

$$|BC| - XI\cos\varphi = E\sin\delta$$

Hence $I\cos\varphi = \frac{E}{X}\sin\delta$

$$|AC| = XI\sin\varphi = E\cos\delta - V$$

Hence $I\sin\varphi = \frac{E}{X}\cos\delta - \frac{V}{X}$

Real power, $P = VI\cos\varphi = \frac{EV}{X}\sin\delta$

$$\therefore P(\delta) = \frac{EV}{X}\sin\delta \;\leftarrow\; \text{power-angle characteristics}$$

δ   is known as the load angle or power angle

Since the real power P depends on the product of phase voltages and the sine of the angle δ between their phasors. In power networks, node voltages must be within a small percentage of their nominal values. Hence such small variations cannot influence the value of real power. Large changes of real power, from negative to positive values, correspond to changes in the sin δ. The system can operate only in that part of the characteristic which is shown by a solid line in Fig. A.1c. The angle

© Springer Nature Singapore Pte Ltd. 2018   159
T. Bambaravanage et al., *Modeling, Simulation, and Control of a Medium-Scale Power System*, Power Systems, https://doi.org/10.1007/978-981-10-4910-1

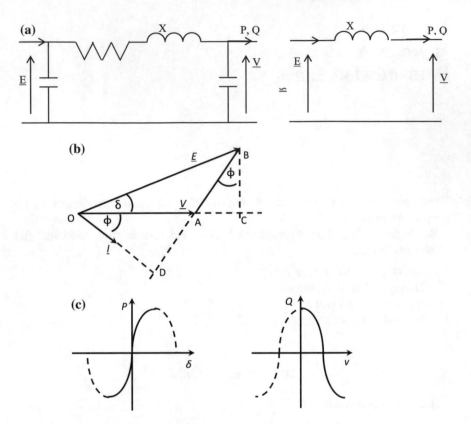

Fig. A.1 a Two-port π equivalent circuit corresponding to an approximated transmission line. **b** Corresponding phasor diagram. **c** Real power and reactive power characteristics

δ is strongly connected with system frequency f; hence the pair 'P and f' is also strongly interrelated.

Reactive power,

$$Q = \frac{EV}{X}\cos\delta - \frac{V^2}{X}$$

$$\cos\delta = \sqrt{1 - \sin^2\delta}$$

$$Q = \sqrt{\left(\frac{EV}{X}\right)^2 - P^2} - \frac{V^2}{X}$$

Due to stability considerations, the system can operate only in that part of the characteristic which is shown by a solid line. The smaller the reactance X, the steeper the parabola; even for small changes in V, cause large changes in reactive power. Obviously the inverse relationship also takes place: a change in reactive power causes a change in voltage.

Hence the three factors that can affect the stability of PS can be identified as:

- Load angle, δ
- Frequency, f
- Nodal voltage magnitude, V

Appendix B
Composite Loads

Usually each composite load represents a relatively large fragment of the system typically comprising

- low- and medium-voltage distribution networks,
- small power sources operating at distribution levels,
- reactive power compensators,
- distribution voltage regulators,
- a large number of different component loads such as motors, lighting and electrical appliances [6].

In the steady state the demand of the composite load depends on the bus-bar voltage V and the system frequency f. The functions describing the dependence of the active and reactive load demand on the voltage and frequency $P(V, f)$ and $Q(V, f)$ are called the *static load characteristics*.

The characteristics $P(V)$ and $Q(V)$, taken at constant frequency, are called the *voltage characteristics* while the characteristics $P(f)$ and $Q(f)$, taken at constant voltage, are called the *frequency characteristics*. The slope of the voltage or frequency characteristic is referred to as the *voltage (or frequency) sensitivity* of the load. Figure B.1, illustrates this concept with respect to voltage sensitivities.

Voltage sensitivities k_{PV} and k_{QV} and the frequency sensitivities k_{PF} and k_{QF} are usually expressed in per units with respect to a given operating point:

$$k_{PV} = \frac{\Delta P/P_0}{\Delta V/V_0}$$

$$k_{QV} = \frac{\Delta Q/Q_0}{\Delta V/V_0}$$

$$k_{Pf} = \frac{\Delta P/P_0}{\Delta f/f_0}$$

© Springer Nature Singapore Pte Ltd. 2018
T. Bambaravanage et al., *Modeling, Simulation, and Control of a Medium-Scale Power System*, Power Systems, https://doi.org/10.1007/978-981-10-4910-1

Fig. B.1 Illustration of the definition of voltage sensitivity

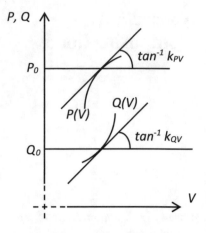

$$k_{Qf} = \frac{\Delta Q/QP_0}{\Delta f/f_0}$$

where,

P_0, Q_0, V_0, f_0, ΔP and ΔQ, are: real power, reactive power, voltage, frequency, real power change, and reactive power change at a given operating point.

A load is considered to be stiff, if at a given operating point, its voltage sensitivities are small.

If,

- $k_{PV} \simeq 0$
  $k_{QV} \simeq 0$, the load is considered to be ideally stiff. The power demand of that load does not depend on the voltage.
- A load is voltage sensitive if k_{PV} and k_{QV} are high
- For a small ΔV change cause high change in the demand, ΔP .
- Usually $k_{PV} < k_{QV}$

Appendix C
Generation Characteristic

In the steady state, the idealized power–speed characteristic of an ith generating unit can be written as:

$$\frac{\Delta\omega}{\omega_n} = -\rho\frac{\Delta P_m}{P_n}\frac{\Delta P_m}{P_n} = -K\frac{\Delta\omega}{\omega_n}$$

$$\frac{\Delta f}{f_n} = -\rho_i\frac{\Delta P_{mi}}{P_{ni}}\frac{\Delta P_{mi}}{P_{ni}} = -K_i\frac{\Delta f}{f_n}$$

In the steady state, all the generating units operate synchronously at the same frequency. When,

$\Delta\omega$ fraction of rated speed
ω_n rated speed
ω *turbine speed*
Δf fraction of frequency
f system frequency
ΔP_T the overall change in the total power generated
N_G no. of generator units
P_m turbine power
P_n nominal power output;

$$\Delta P_T = \sum_{i=1}^{N_G} \Delta P_{mi}$$

$$-K_i\frac{\Delta f}{f_n} = \frac{\Delta P_{mi}}{P_{ni}}$$

© Springer Nature Singapore Pte Ltd. 2018
T. Bambaravanage et al., *Modeling, Simulation, and Control of a Medium-Scale Power System*, Power Systems, https://doi.org/10.1007/978-981-10-4910-1

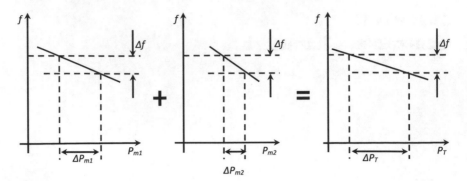

Fig. C.1 Generation characteristic as the sum of speed–droop characteristics of all the generation units

$$\therefore \Delta P_T = -\frac{\Delta f}{f_n} \sum_{i=1}^{N_G} K_i P_{ni}$$

$$\therefore \Delta P_T = -\Delta f \sum_{i=1}^{N_G} \left(\frac{K_i P_{ni}}{f_n}\right) \tag{C.1}$$

Figure C.1, illustrates how the characteristics of individual generating units can be added according to Eq. (C.1) to obtain the equivalent generation characteristic. This characteristic defines the ability of the system to compensate for a power imbalance at a situation of a system frequency deviation from its rated value. For a power system with a large number of generating units, the generation characteristic is almost horizontal such that even a relatively large power change only results in a very small frequency deviation. This is one of the benefits due to combining generating units into one large system.

To obtain the equivalent generation characteristic of Fig. C.2, it has been assumed that the speed–droop characteristics of the individual turbine-generator units are linear over the full range of power and frequency variations. In practice the output power of each turbine is limited by its technical parameters. The speed–droop characteristics of a turbine with an upper limit is shown in Fig. C.2.

If a turbine is operating at its upper power limit then a decrease in the system frequency will not produce a corresponding increase in its power output. At the limit $\rho = \infty$ or $K = 0$ and the turbine does not contribute to the equivalent system characteristic. Consequently the generation characteristic of the system will be dependent on the number of units operating away from their limit at part load; that is, it will depend on the spinning reserve, where **the spinning reserve is the difference between the sum of the power ratings of all the operating units and their actual load**.

Fig. C.2 Speed–droop characteristic of a turbine with an upper limit

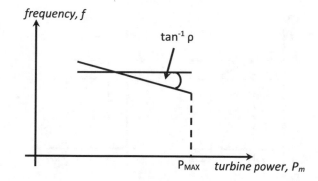

The allocation of spinning reserve is an important factor in power system operation as it determines the shape of the generation characteristic. This is demonstrated in Fig. C.3, with two generating units.

In Fig. C.3a, the spinning reserve is allocated proportionally to both units (which operate at a frequency of f_0) and the maximum power of both generators is reached at the same operating frequency f_1. The sum of both characteristics is then a straight line (as given in Eq. (C.1)), up to the maximum power $P_{MAX} = P_{MAX\,1} + P_{MAX\,2}$.

Figure C.3b shows a situation where the total system reserve is the same (equal to the amount of the previous case), but it is allocated solely to the second generator. That generator is loaded up to its maximum at the operating point (frequency f_2). The resulting total generation characteristic is nonlinear and consists of two lines of different slopes. The first line is formed by adding both inverse droops, $K_{T1} \neq 0$ and $K_{T2} \neq 0$, in Eq. (C.3). The second line is formed noting that the first generator operates at maximum load and $K_{T1} = 0$, so that only $K_{T2} \neq 0$ appears in the sum in Eq. (C.3). Hence the slope of that characteristic is higher (Fig. C.3).

Generally, the number of units operating in a real system is large. Some of them are loaded to the maximum but others are partly loaded, generally in a non-uniform way, to maintain a spinning reserve. Adding up all the individual characteristics would give a nonlinear resulting characteristic consisting of short segments with increasingly steeper slopes. That characteristic can be approximated by a curve as shown in Fig. C.4. The higher the system load, the higher the droop until it becomes infinite $\rho_T = \infty$, and its inverse $K_T = 0$, when the maximum power P_{MAX} is reached. If the dependence of a power station's auxiliary requirements on frequency were neglected, that part of the characteristic would be vertical (shown as a dashed line in Fig. C.4).

The total system power generation is equal to the total system load (P_L), including transmission losses.

$$\sum_{i=1}^{N_G} P_{mi} = P_L$$

(a)

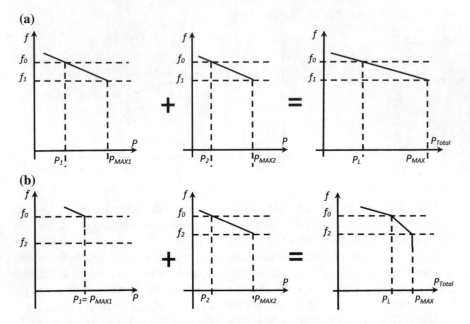

(b)

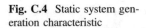

Fig. C.3 Influence of the turbine upper power limit and the spinning reserve allocation on the generation characteristic

Fig. C.4 Static system generation characteristic

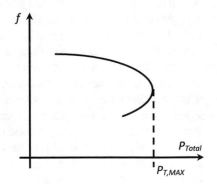

Equation (C.1)/P_L gives:

$$\frac{\Delta P_T}{P_L} = -K_T \frac{\Delta f}{f_n} \; or \; \frac{\Delta f}{f_n} = -\rho_T \frac{\Delta P_T}{P_L} \tag{C.2}$$

where,

$$K_T = \frac{\sum_{i=1}^{N_G} (K_i P_{ni})}{P_L} \tag{C.3}$$

$$\rho_T = \frac{1}{K_T}$$

Equation (C.2) describes the linear approximation of the generation characteristic calculated for a given total system demand. Further, the coefficients in Eq. (C.3) are calculated with respect to the total demand, not the sum of the power ratings, so that ρ_T is the local speed-droop, of the generation characteristic and depends on the spinning reserve and its allocation in the system as demonstrated in Fig. C.4.

Appendix D

Generation and transmission network of Sri Lanka as at 2011.

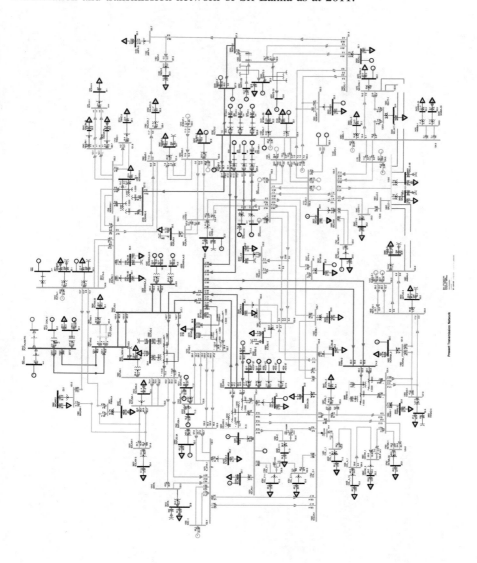

© Springer Nature Singapore Pte Ltd. 2018
T. Bambaravanage et al., *Modeling, Simulation, and Control of a Medium-Scale Power System*, Power Systems, https://doi.org/10.1007/978-981-10-4910-1

References

1. R.C. Dugan, M.F. McGranaghan, S. Santoso, H.W. Beaty, *Electrical Power Systems Quality*, 2nd edn. (McGraw-Hill, 2004).
2. PUCSL, Investigation Report on Power System Failures on 9th October 2009, Public Utilities Commision of Sri Lanka, 2010 February.
3. J. Barkans, D. Zalostiba, *Protection Against Black-outs and Self-restoration of Power Systems* (RTU Publishing House, Riga, 2009)
4. P.M. Anderson, *Power System Protection* (IEEE press, Wiley-Interscience, 1999)
5. Seimens, Technical assesment of Sri Lanka's renewable resource based electricity generation, Final Report, RERED Project, Project No. 108763-0001 (Seimens Power Technologies International Ltd, England, UK, 2005).
6. J. Machowski, J.W. Bialek, J.R. Bumby, *Power System Dynamics, Stability and Control* (Wiley, New York, 2008)
7. P. Kundur, J. Paserba, V. Ajjarapu, G. Andersson, Definition and classification of power system stability, IEEE/CIGRE joint task force on stability terms and definitions. IEEE Trans. Power Syst. **19**(2), 1387–1401 (2004).
8. M.Q. Ahsan, A.H. Chowdhury, S.S. Ahmed, I.H. Bhuyan, M.A. Haque, H. Rahman, Technique to develop auto load shedding and islanding scheme to prevent power system blackout. IEEE Trans. Power Syst. **27**(1), 198–205 (2012).
9. E. Vaahedi, *Practical Power System Operation* (IEEE Press, Wiley, New York, 2014)
10. M. Giroletti, M. Farina, R. Scattolini, A hybrid frequency/power based method for industrial load shedding. Elsevier Int. J. Electr. Power Energy Syst. **35**, 194–200 (2012)
11. P. Kundur, *Power System Stability and Control* (McGraw-Hill Inc, New York, 1993)
12. H. You, V. Vittal, J. Jung, C. Liu, M. Amin, R. Adapa, An Intelligent adaptive load shedding scheme, in *14th PSCC*, Sevilla, Spain, 24–28 June, 2002.
13. A.J. Wood, B.F. Wollenberg, G.B. Sheble, *Power Generation, Operation, and Control*, 3rd edn. (IEEE Press, Wiley, Hoboken, NJ, 2014)
14. F. Saccomanno, *Electric Power Systems, Analysis and Control* (IEEE press, Wiley-Interscience, 2003).
15. M. Eremia, M. Shahidehpour, *Hand book of Electrical Power System Dynamics: Modeling, Stability and Control* (IEEE Press, Wiley, 2013).
16. *Guide for grid interconnection of embedded generators, Part I: Application, evaluation and interconnection procedure,* Ceylon Electricity Board, December, 2000.
17. Wikipedia, https://en.wikipedia.org/wiki/Photovoltaic_system. Accessed 2 Oct 2015.
18. *Guide for grid interconnection of embedded generators, Part 2: Protection and operation of grid interconnection,* Ceylon Electricity Board, December, 2000.
19. CEB, Long term transmission development plan 2011–2020. Ceylon Electricity Board, July, 2011.

© Springer Nature Singapore Pte Ltd. 2018   173
T. Bambaravanage et al., *Modeling, Simulation, and Control of a Medium-Scale Power System*, Power Systems, https://doi.org/10.1007/978-981-10-4910-1

20. J. Liang, G. Venayagamoorthy, R. Harley, Wide-area measurement based dynamic stochastic optimal power flow control for smart grids with high variability and uncertainty. IEEEE Trans. Smart Grid **3**(1), 59–69 (2012).
21. CEB, http://www.ceb.lk/downloads/st_rep/stat2010.pdf. Ceylon Electricity Board. Accessed 12 Oct 2015.
22. CEB, http://www.ceb.lk/downloads/st_rep/stat2013.pdf. Ceylon Electricity Board. Accessed 12 Oct 2015.
23. T. Bambaravanage, A. Rodrigo, S. Kumarawadu, N.W.A. Lidula, Under-frequency load shedding for power systems with high variability and uncertainty, in *ISPCC 2013 Proceedings of the IEEE international conferrence on Signal Processing, Computing and Control*, Solan, India, 2013.
24. ABB, *Load Shedding Controller, PML630 Product Guide,* Vaasa, Finland: ABB, 2011.
25. P. Mahat, Z. Chen, B. Bak-Jensen, Under frequency load shedding for an islanded distribution system with distributed generators. IEEE Trans. Power Deliv. **25**(2), 911–918 (2010).
26. V.V. Terzija, Adaptive under-frequency load sheddig based on the magnitude of the disturbance estimation. IEEE Trans. Power Syst. **21**(3), 1260–1266 (2006).
27. M. Begovic, D. Novose, D. Karlsson, C. Henville, G. Michel, Wide-area protection and emergency control. Proc. IEEE **93**(5), 876–891 (2005).
28. B. Delfino, S. Massucco, A. Morini, P. Scalera, F. Silvestro, Implementation and comparison of different under frequency load-shedding schemes, in *IEEE Power Engineering Society Summer Meeting,* 2001, pp. 307–312, 2001.
29. P. Anderson, M. Mirheydar, An adaptive method for setting under-frequency load shedding relays. IEEE Trans. Power Syst. **7**(2), 647–655 (1992).
30. M. Gunawardena, C. Hapuarachchi, D. Haputhanthri, I. Harshana, Capacity limit of the single largest generator uni, to maintain power system stability through a load shedding program. Department of Electrical Engineering, University of Moratuwa, December 2011.
31. J.R. Jones, W.D. Kirkland, Computer algorithm for selection of frequency relays for load shedding. IEEE Comput. Appl. Power **1**(1), 21–25 (1988).
32. H. Bentarzi, A. Quadi, N. Ghout, F. Maamri, N.E. Mastorakis, A new approach applied to adaptive centralized load shedding scheme, in *CSECS'09 Proceedings of the 8th WSEAS International Conference on Circuits, Systems, Electronics, Control & Signal Processing,* Wisconsin, USA, 2009.
33. K. Wong, B. Lau, Algorithm for load-shedding operations in reduced generation periods. IEEE Proc. **139**(6), 478–490 (1992).
34. R. Maliszewski, R. Dunlop, G. Wilson, Frequency actuated load shedding and restoration, part I—philosophy. IEEE Trans. Power App. Syst. **PAS-90**(4), 1452–1459 (1971).
35. Network protection & automation guide: protective relays, measurements & control, Alstom Grid, May, 2011.
36. M.G. Simoes, B. Palle, S. Chakraborty, C. Uriarte, Electrical model development and validation for distributed resources. National renewable energy laboratory, Colorado, USA, April 2007.
37. C. Muller, *User's guide on the use of PSCAD* (Manitoba HVDC Research Centre, Manitoba, 2010)
38. P. Wilson, *User's guide: a comprehensive resource for EMTDC* (Manitoba HVDC Reseaech Centre Inc., Manitoba, 2005)
39. Introduction to PSCAD/EMTDC (Manitoba HVDC Research Centre Inc., Manitoba, 2003).
40. Y.S.M.I. Xi-Fan Wang, *Modern Power System Analysis* (Springer Science+Business Media, LLC, NY, USA, 2008).
41. ECB, Long term transmission development studies 2005–2014. Ceylon Electricity Board.
42. Nexans, 60–500 kV High Voltage under ground power cables: XLPE insulated cables (Nexans, Paris, France).

43. G.F. Moor,*Electric cables handbook-Third Edition: BICC Cables* (Blackwell Science, Oxford, UK, 1997)
44. E.B. Joffe, K.-S. Lock,*Grounds for Drounding: A Circuit-to-System Handbook* (IEEE Press; Wiley, NJ, USA, 2010)
45. T. Wildi, Chapter 10; Practical transformers, in *Electrical Machines, Drives and Power Systems*, 5th edn. (Prentice Hall, New Jersey, 2002), pp. 197–224
46. A.V. Meier, Generators, in *Electric Power Systems: A Conceptual Introduction* (IEEE Press, Wiley-Interscience, New Jersey, USA, 2006), Chapter 4, pp. 85–126.
47. S.W. Smith, in *The scientist and Engineer's Guide to Digital Signal Processing,* 2nd edn. (California Technical Publishing, California, USA, 1999), pp. 35–66.
48. K. Ogata, *Modern Control Engineering*, 4th edn. (Prentice-Hall, New Jersey, 2002)
49. CEB, Long term transmission development plan 2011–2020. Ceylon Electricity Board, July, 2011.
50. PUCSL, Generation performance in Sri Lanka 2014. Public utilities commission of Sri Lanka, 2014.
51 CEB, Generator Interconnection of Sri Lanka. Ceylon Electricity Board.
52 C. Huang, S. Huang, A time-based load shedding protection for isolated power systems. Electric Power Systems Research **52**, 161–169 (1999)
53 M. Eremia, M. Shahidehpour, *Handbook of Electrical Power System Dynamics: Modeling, Stability and Control* (IEEE Press, Wiley, New York, 2013)
54 SIEMENS, *SIPROTEC Over Current Time Protection 7SJ80; V4.6 manual.,* SIEMENS.
55 M.Q. Ahsan, A.H. Chowdhury, S.S. Ahmed, I.H. Bhuyan, Technique to develop auto load shedding and islanding scheme to prevent power system blackout. IEEE Trans. Power Syst. **27**(1), 198–205 (2012).
56 http://en.wikipedia.org/wiki/Photovoltaic_system. Accessed 2 Oct 2015.
57 CEB, http://www.ceb.lk/downloads/st_rep/stat2011.pdf. Ceylon Electricity Board. Accessed 12 October 2015.
58 CEB, http://www.ceb.lk/downloads/st_rep/stat2012.pdf. Ceylon Electricity Board. Accessed 12 Oct 2015.
59 H. Bevrani, A.G. Tikdari, T. Hiyama, Power system load shedding: key issues and new perspectives. World Academy of Science, Engineering and Technology **65**, 177–182 (2010)
60 J. Ford, H. Bevrani, G. Ledwich, Adaptive load shedding and regional protection. Elsevier Int. J. Electr. Power Energy Syst. **31**, 611–618 (2009)
61 S. Arnborg, G. Andersson, D.J. Hill, I.A. Hiskens, On under voltage load shedding in power systems. Electr. Power Syst. **19**(2), 141–149 (1997).
62 CEB, Long term Generation expansion plan 2015–2034. Ceylon Electricity Board, July 2015.
63 http://www.most.gov.mm/techuni/media/EP_03041_4.pdf. Accessed 10 March 2013.
64 P. Mahat, Z. Chen, B. Bak-Jensen, Control and Operation of distributed generation in distribution systems. ScienceDirect, Electr. Power Syst. Res. **81**, 495–502 (2011)
65 T. Bambaravanage, S.K.A. Rodrigo, N.W.A. Lidula, A new scheme of under frequency load shedding and islanding operation. Annual Trans. IESL**1** (Part B), 290–296 (2013).
66 T. Bambaravanage, S.K.A. Rodrigo, Comparison of three Under-Frequency Load Shedding Schemes referring to the Power System of Sri Lanka. Engineer: J Inst Eng Sri Lanka **49** (1):41 (2016)

Printed in the United States
By Bookmasters